T0314171

TROUBLESOME SCIENCE

RACE, INEQUALITY, AND HEALTH

RACE, INEQUALITY, AND HEALTH
Edited by Samuel Kelton Roberts Jr. and Michael Yudell

The Race, Inequality, and Health series explores how race has become a basis for discrimination and inequality, as well as a foundation for reactionary or affirmative group politics. The goal of this series is to offer an intellectual space for leading scholars, combining work in history, the social sciences, the biological sciences, and public health that will deepen our understanding of how ideological and scientific claims about race and race difference have impacted health and society both historically and in the present day.

Michael Yudell, *Race Unmasked: Biology and Race in the Twentieth Century*

TROUBLESOME SCIENCE

· · · · · · · · · · · · · · · · ·

The Misuse of Genetics and Genomics in
Understanding Race

ROB DESALLE AND
IAN TATTERSALL

COLUMBIA UNIVERSITY PRESS

New York

Columbia University Press
Publishers Since 1893
New York   Chichester, West Sussex
cup.columbia.edu

Library of Congress Cataloging-in-Publication Data
Names: DeSalle, Rob, author. | Tattersall, Ian, author.
Title: Troublesome science : the misuse of genetics and genomics
in understanding race / Rob DeSalle and Ian Tattersall.
Description: New York : Columbia University Press, [2018] |
Includes bibliographical references and index.
Identifiers: LCCN 2017052052 | ISBN 9780231185721 (cloth : alk. paper) |
ISBN 9780231546300 (e-book)
Subjects: | MESH: Genetics, Population | Continental Population Groups |
Genomics | Biological Evolution
Classification: LCC QH455 | NLM QU 450 | DDC 576.5/8—dc23
LC record available at https://lccn.loc.gov/2017052052

Columbia University Press books are printed on permanent
and durable acid-free paper.
Printed in the United States of America

Cover image: Byron Kim, *Synecdoche* (1991–present).
Courtesy of the National Gallery of Art

Cover design: Chang Jae Lee

To the memory of our colleague and good friend
Bob Sussman, who detested inequality

Contents

Preface

In a perfect world, there would be no need for another book on science and race. After all, it has been repeatedly demonstrated that attempting to subdivide our species *Homo sapiens* into the objectively recognizable units that scientific analysis requires is an entirely hopeless task. Indeed, we thought that we had reasonably effectively shown this ourselves a few years ago, in our 2011 book *Race? Debunking a Scientific Myth*. Our late and hugely lamented colleague Robert W. Sussman subsequently complemented our efforts to what should have been great effect in his *The Myth of Race: The Troubling Persistence of an Unscientific Idea*, in which he used eugenics as a lens through which to elegantly explain how bigotry so often contrives to be mistaken for science.

Sadly, though, even a cursory review of the recent literature is enough to demonstrate the tenacity of the belief that the real existence of human races is somehow justified by biological science. Comparative molecular biologists, for example, continue to "stratify" their geographic samples as a prelude to further study, while many clinical pharmacological studies carry on the venerable tradition of classifying their subjects by race. The U.S. government persists in requesting its citizens pigeonhole themselves by race for a whole host of different purposes; and equally disturbing is the ongoing success of popular books such as Nicholas Wade's *A Troublesome Inheritance: Genes, Race and Human History*, which purports to demonstrate the objective existence of human races on scientific grounds.

Clearly, widespread confusion still exists about the nature of the evidence on which the human species is subdivided and about the objectivity or applicability of the scientific methods used to do this. Most of those methods belong to taxonomy, the science devoted to the ordering and classification of the riotous diversity of the natural world of which we form a part. As taxonomists ourselves, we fully appreciate the power of such methods to help us organize the complexities of a biosphere with a long and branching evolutionary history. But we equally recognize that these methods are totally unsuited to classifying variants within species, the fundamental units of analysis when we examine the structure of life. For while species are individuated historical units, off on their own unique evolutionary trajectories, variants within them are not only in principle ephemera, but almost by definition have no discrete boundaries. Consequently, the procedures that have been developed for ordering different species cannot usefully be applied to the undeniable variety that exists within *Homo sapiens*.

Hence this book, in which we hope to demonstrate this simple reality by looking in detail at just how taxonomists go about their job, and at how the techniques they use both to recognize species, and to evaluate the relationships among them, stumble badly when misapplied to the study of human variety.

We begin by examining the evolutionary process that produced us, and how it may be represented via visual metaphor. We then proceed to a consideration of what species are and how they may be recognized, organized into phylogenetic trees, and named. The first modern techniques we look at are DNA barcoding and profiling, following which we anchor our discussion in the human context by looking briefly at early notions of divergence within the human species. We follow this with an accounts of the discovery of human blood groups and of the "mitochondrial Eve" and "Y-chromosome Adam" concepts that ultimately derived from that breakthrough, before we discuss the remarkable technologies that have allowed us to learn the genomic sequences of some of our extinct relatives.

With an understanding of the genomics involved, we are able to next explore what the evidence of the genes can tell us about the ancient migrations of human populations and about how *Homo sapiens* contrived to take over the entire habitable world in a remarkably short span of time. Once this is explained, we are in a better position to scrutinize the methods that have been specifically applied in recent years to the results of those ancient population migrations, which naturally enough resulted in extensive demographic and genomic mixing. We demonstrate the inadequacy of clustering methods in sorting out the resultant cline-filled, infraspecies variety and reveal the similar limitations inherent in the more recently introduced STRUCTURE analyses of human populations.

Finally, we examine the claims made in Nicholas Wade's book in some detail, concluding that his assertions fail as science in at least seven distinct ways. At the end of this methodological journey, we trust that we will have convinced the reader that the concept of identifiable human "races" has neither utility nor scientific credibility in understanding either human demographic history or the current genomic variety of *Homo sapiens*. In biological terms, "races" as definable entities simply do not exist, and none of the taxonomic and systematic approaches that have been applied to the issue are suited to demonstrating that they do.

On the other hand, there is no denying that in both the cultural and political spheres, the concept of "race" remains extremely potent and continues to blight both social equality and ethnic relationships within American society. Accordingly, we conclude our survey with an epilogue that briefly considers the social aspects of race and the auguries for the future, always remembering that the naming or defining of any human subgroup has the potential to produce a destructive "us versus them" reaction. Discouragingly, we find only modest grounds for optimism in this respect, at least for the decades immediately to come; but there is no doubt in our minds that the prospects for amelioration will be vastly improved if researchers, journalists, and the public somehow find themselves able to rid themselves of the fallacy that "race" means anything in terms of science. We offer this book in the hope that it will contribute, however infinitesimally, toward the realization of that goal.

Acknowledgments

We thank Michael Yudell, coeditor of the series in which this book appears, for the invitation to participate in this important initiative. At Columbia University Press we would never have been able to proceed without the enthusiastic support and encouragement of our editor Patrick Fitzgerald and his associates Ryan Groendyk and Brian Smith, active participants in this project. We are also deeply indebted to Kathryn Jorge at CUP, and to Ben Kolstad at Cenveo, for their meticulous and expert attention during the production process. At the American Museum of Natural History, we express our appreciation to Apurva Narechania, Martine Zilversmit, Jeff Rosenfeld, and Michael Tessler for their help with phylogenetic analysis and to Vivian Schwartz for her able editorial guidance. As always, we are both far more than grateful to Patricia Wynne for the wonderful illustrations that impart life to our text. Finally, Erin DeSalle and Jeanne Kelly consistently showed imperturbable good humor during the writing phase; our deepest thanks, as ever, to you both.

TROUBLESOME SCIENCE

1

• • • •

Evolutionary Lessons

Because all life on our planet shares a common ancestor, there is an unbroken continuity among the genomes of all living things on Earth, including us. That same common ancestry also means that every species living on our planet today is the product of well over three billion years of evolution. Again, this includes us. But while human beings are incontestably the outcome of an unimaginably long biological history, our species *Homo sapiens* is a newcomer. It appeared as an anatomically distinctive entity a mere two hundred thousand years ago, and its members began behaving in our unusual modern manner more recently yet. This short history matters, because it means that all the physical variation we perceive within our species is of extremely recent origin. Indeed, in evolutionary terms that variation is, in a very real sense, epiphenomenal. Looking around on any busy street in a major city, many members of our egocentric—even narcissistic—species might find this claim overstated, or even absurd; but in the larger scheme of things, the physical variations we so clearly perceive within *Homo sapiens* are minor indeed—though this reality can only be properly understood in the context of the grand evolutionary process that has molded ours and every other species on the planet. To comprehend what this process tells us about ourselves, we need to begin with the nineteenth-century researches of two men who literally revolutionized our understanding of our own place in nature: Charles Darwin and Alfred Russel Wallace.

The story is well known of how these two brilliant biologists almost simultaneously propounded the notion that the riotous diversity of the living world is the product of evolution; and while we should never forget Wallace's contribution, we will focus in this chapter on Darwin's seminal formulation of the issue. Darwin was a man of extraordinarily wide-ranging curiosity and imagination, and in a long and industrious career he did many things that ended up being foundational in natural history. He was a fount of important ideas, and several of them are particularly important for our understanding of evolution.

One of these was perhaps as much a way of chronicling and organizing observations as it was an idea. Throughout his career as a naturalist, Darwin was an astute observer and notetaker: a habit firmly established by 1831 when, at the tender age of twenty-two, he departed on his round-the-world voyage as an unofficial naturalist on the Royal Navy sloop HMS *Beagle*. As he made more and more observations, took more and more notes, and collected more and more specimens during the long journey, Darwin became increasingly impressed by the organization of living creatures into what today we would call "nested sets." Each species belonged, with others like it, in a larger grouping known as the "genus." Genera in turn belonged to larger "families," which could be grouped into "orders," and so on up the line until finally all living things were embraced by a giant unit whose members simply shared life itself. And it was Darwin's genius to recognize, almost certainly well before his five-year odyssey ended, that only one mechanism could have given rise to a pattern like this. As he later very neatly put it, this was "descent with modification." All life was ultimately descended from a single progenitor species whose descendants had diversified over the eons, much as members of the various branches of a human family might ultimately become bakers or sailors or financiers. Inconveniently, though, this key realization inevitably breached one of the great received wisdoms of the day: namely, that species remained essentially fixed as the Creator had made them (by one estimate, in 4004 BC). For in Darwin's scheme, species inevitably had to "transmute." In other words, they had to be able to give rise to offspring that were not exactly like themselves.

Darwin was not the first to propose such a thing. Eighteenth-century figures such as the American cleric Samuel Stanhope Smith had already hinted, specifically in the context of human races, that adaptation might be involved in the generation of diversity; and Darwin's own grandfather, the physician Erasmus Darwin, had speculated toward the end of the century that each organism might be capable of "continuing to improve by its own inherent activity, and of delivering down those improvements by generation to its posterity."

Among mainstream biologists, the French naturalist Jean-Baptiste Lamarck had suggested back in 1809 (the year of Darwin's birth) that fossil mollusks found in the Paris Basin fell into lineages that steadily modified over time; and the Italian geologist Giambattista Brocchi soon followed up in 1814 with the notion that, while the fossil species he had collected in Italy's Apennine Mountains typically spent their sometimes extended lifetimes essentially unchanged, they also gave rise to modified descendants. Between them, Lamarck and Brocchi had unearthed the key elements of our modern understanding of evolution as early as the first quarter of the nineteenth century, but these two breathtakingly original thinkers were far ahead of their time, and Lamarck was ridiculed while Brocchi was forgotten. Darwin, in contrast, appeared on the scene at a time when the geologists had pretty much nailed down the scientific case for believing that Earth had a very long history indeed (providing plenty of time in which change could occur); and he additionally came up with a very ingenious explanation for how that change could have come about.

Darwin's chosen mechanism for evolution involved what he called natural selection. Both Wallace and Darwin had read and taken inspiration from Thomas Malthus's publication in 1798 of *An Essay on the Principle of Population*, in which the English scholar had argued that populations tended inevitably to outgrow the resources available to them. The argument for natural selection built on this, based on the realization that all members of any species' population vary in their inherited physical attributes and that many more individuals were born in every generation than would ever reproduce. It was the "fittest" individuals—those whose inherited traits best adapted them to prevailing circumstances—who would survive and most successfully reproduce. Those less well adapted would fall by the wayside, taking their inferior adaptations with them. In this way, the species' lineage would change insensibly from generation to generation, steadily tending to a better-adapted state. This was a magnificently reductionist formulation with instant appeal to members of a species that craves neat explanations, and although, as we will see, it is deeply flawed, it ultimately proved to be evolution's most powerful selling point.

The third major advance that Darwin made was in changing how we recognize entities in nature. The ornithologist and evolutionary biologist Ernst Mayr pointed out that Darwin instituted what he called "population thinking." The prevailing way of looking at the natural world when Darwin was learning his natural history was derived from the typological thinking of the ancient Greeks. This focused on the "essence," or average, of each kind of organism and considered variation among individuals as a more or less philosophical construct. While observable, variation had to be abstracted away so that the "type" could be described. This way of thinking about the natural world has its uses, as we will see in subsequent chapters; but it stood in the way of the understanding of nature that Darwin developed in 1859 in his great book *On the Origin of Species by Means of Natural Selection; or, the Preservation of Favoured Races in the Struggle for Life* (hereafter referred to as "the *Origin*"). This is because population thinking looked at the natural world and saw the variation that natural selection requires, while the average or mean was an abstraction that inherently limited natural historians' understanding of the natural world. This break from typological thinking will become important when we discuss the population biology of our species; for without it, we are at a loss to describe its incredibly complex history.

The fourth major advance that Darwin made lay in articulating a clearcut way to view evolutionary patterns. While he wasn't the first to suggest that branching diagrams could be used to represent the relationships of organisms, Darwin employed them in the context of descent with modification. Three bits of evidence suggest that he saw early on the utility of evolutionary trees in describing descent with modification. The first is his "I think" page in a notebook dating to 1837. The diagram on this page is now famous, and in its tattooed form adorns the person of many an evolutionary biologist. It shows a branching diagram above which Darwin has scrawled "I think" (figure 1.1). The diagram is accompanied by a paragraph in which Darwin explains the meaning of its tips and nodes.

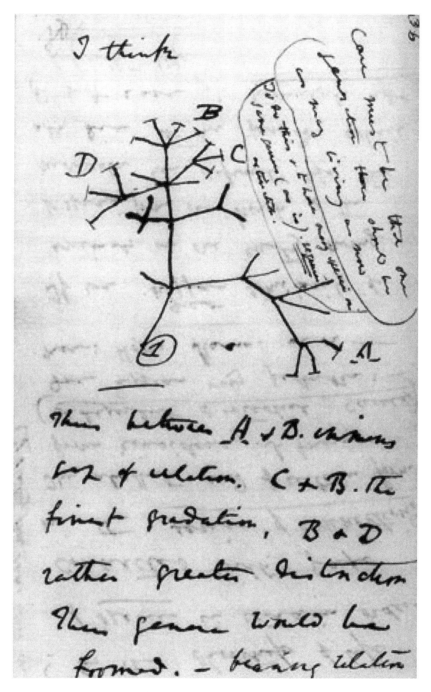

Figure 1.1 The famous "I think" figure from Darwin's notebooks. Darwin describes the figure in the writing below the drawing and illustrates the meaning of its tips and nodes. He labeled four tips, "A" through "D," as species, scrawling next to the drawing: "Thus between A & B immense gap of relation. C & B the finest gradation, B & D rather greater distinction."

We are a bit puzzled that Darwin might have tricked himself a bit in his comparison of "B" and "C" to "B" and "D." If one looks at the diagram closely, "B" is actually just as close to "C" as it is to "D." There are the same number of ancestors between "B" and "C" as there are between "B" and "D."

He labeled four tips, "A" through "D," as species, scrawling next to the drawing: "Thus between A & B immense gap of relation. C & B the finest gradation, B & D rather greater distinction." In looking at the diagram, we note that he is correct about "B" and "C" being more closely related to each other than "B" and "A," and that "B" and "A" have an "immense" gap between them relative to the "finest gradation" between "B" and "C."

Darwin's second and third uses of branching diagrams come from the *Origin*. Indeed, his famous hierarchical branching diagram (figure 1.2) is the only figure in the book's first edition, and he takes up nearly one-third of his chapter on natural selection with a discussion of the diagram and its role in representing descent with modification. Referring directly to the diagram, Darwin declares, "Thus, as I believe, species are multiplied and genera are formed." Once again, he is moving us into new intellectual territory with his use of a branching diagram (a "tree") to explain the heart of the thinking in his book.

The third bit of evidence for his preference for this new "tree thinking" is found in the beautiful prose that he uses in the *Origin* to describe what he calls "the great Tree of Life, which fills with its dead and broken branches the crust of the earth, and covers the surface with its ever-branching and

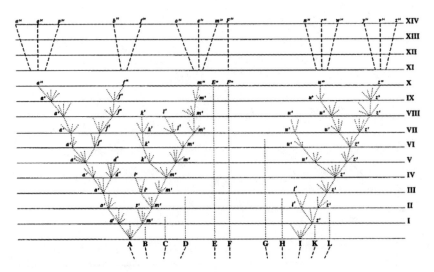

Figure 1.2 The only figure in the *Origin*. Darwin used nearly one-third of his chapter on natural selection (chapter 4) in discussing this diagram.

beautiful ramifications" (Darwin 1859, 82). This preference for tree thinking was a revolutionary advance in understanding the structure of the living world and is one to which we will return repeatedly in this book.

Since Darwin's time, thousands of people have dedicated their lives to the study of nature in the context that he created and have built on his accomplishment. The *Origin* created a context for all future studies of the natural world; but naturally enough, given the complexity of the biosphere, the full story of how evolution functions is still being worked out. Nevertheless, there are some generalities about which we can be pretty confident. The first of these draws on population thinking when it states that it is *populations* that evolve. Individuals do not. Individuals live and die; and while their cells—and perhaps even their genomes—change during their lifetimes, they don't evolve. This means that we need to be careful when we are talking about this or that individual in an evolutionary context. Another important generality is that natural selection is not the only force that can change populations; and accordingly, in the 1960s, researchers began to realize that some genetic innovations in populations were neither beneficial nor detrimental to the populations concerned. What this "neutral" theory of evolution means is that a great deal of the evolution going on in populations is unrelated to adaptation or function.

Neutrality leads to yet another generality about evolution, namely that in small populations strange things can happen that have little or nothing to do with natural selection. The effect here is a probabilistic one, so you can look at it in the following way. If we were to bet you that we could flip 100 heads in a row, you should take that bet every time, for the probability of this happening would be $1/2^{100}$ or 1 in $8(10^{31})$: a very small number indeed. On the other hand, if we wanted to wager you that we could roll two heads in a row you might be less ready to take the bet, since the probability of this event is just $1/2^2$ or 0.25. In the same way, small sample sizes may allow for events that otherwise seem improbable. This has surely been an important factor in human evolution, since only recently have our population sizes become large enough to be considered panmictic (randomly mating). As a result, we cannot simply say that natural selection is the only driving factor of evolution, and instead must include as significant influences those chance factors ("genetic drift") that occur because of small sample sizes.

Noting that researchers in evolutionary biology have in the past made mistakes about how the evolutionary process works does not mean that evolution itself is a tarnished or rejectable idea. Quite the opposite; all good science proceeds by constant improvement. We can point to the study of gravity as an example. The first real theory of gravity (Newton's) was superseded by relativity, but at no time during the development of a cohesive theory of gravity did gravity itself stop working. Similarly, while the theory of evolution has been challenged, refined, and retooled since Darwin and Wallace originally placed natural selection at the heart of the phenomenon, at no time during this development has evolution ceased to be a factor in change on our planet. Science is not a search for truth; rather, it is a matter of constantly refining our views of how the world works by rejecting ideas that turn out to be wrong; and it will always be a work in progress.

As we pointed out earlier, the notion of genetic drift was developed to explain chance events that natural selection has little if anything to do with. Soon thereafter, evolutionary biologists began realigning their view of nature, and of the role of natural selection in the variation and diversity they observed around them. In the early 1970s the Harvard scientists Stephen Jay Gould and Richard Lewontin challenged the prevailing paradigm that evolution proceeds as a function of adaptation. They claimed that evolutionary biologists and natural historians had fallen into a rut of relying on natural selection and adaptation as explanations for almost everything they observed in nature. To show that this "adaptationist program" was flawed, they pointed to four reasons why evolution did not strive toward the perfection which was the logical outcome of adaptation alone.

First, if variants in populations can evolve regardless of selection, then not all change will be adaptive. Second, many of the phenotypes we observe in nature—and are tempted to regard as adaptive—turn out actually to be compromises. Take, for instance, human knees. Our knees have evolved to support us as an upright animal walking on two legs; and indeed, our turn to bipedalism has been viewed by many as one of the most important events in our evolutionary history. Darwin himself introduced this notion in his book *The Descent of Man*, when he stated: "The hands and arms could hardly have become perfect enough to have manufactured weapons, or to have hurled stones and spears with a true aim, as long as they were habitually used for locomotion" (Darwin 1871, 35). Note his use of the word "perfect," which is where much of the emphasis on adaptation and perfection in evolution in the early part of the twentieth century originated.

But our poor knees! They are hardly perfect at all, as the hundreds of thousands of people each year who need knee surgery—either from hyperextending the bones and muscles of the knee by accident or by simple overuse—can attest. The evolutionary step to bipedalism turns out to have been a compromise in which the benefits from walking upright were counterbalanced by the negative impact on our knee joints (not to even mention the strain on our hips and spines).

Third, suggesting that evolution strives toward perfection and is always adaptive ignores the very processes that Darwin outlined as necessary for evolution to proceed. Remember that Darwin presented variation as the currency of evolutionary change. If there is a perfect solution to an evolutionary problem, and the variation that might lead to that solution does not happen to exist, the process will settle for whatever is there. There is no way to conjure up the variation that will lead to perfection.

This reflects the fact that a lot of evolution is contingent on what has already evolved. And indeed, a trait that might appear to be a perfect solution for a particular evolutionary problem might have evolved for completely unrelated reasons. Gould and Lewontin called such structures "spandrels," using the beautiful dome structures of Saint Mark's Cathedral in Venice to make their point. Ceiling paintings fit beautifully and "perfectly" into the spandrels inside the domes, and one could make a myopic argument that the spandrels were constructed to hold the artwork. But the spandrels are only there, of course, as by-products of the arches that hold the massive domes in place. The artwork within the spandrels is an afterthought that has nothing to do with the structure of the domes.

None of this is to suggest that natural selection is unimportant in evolution. Indeed, it is a mathematical certainty that it will be operating in any population in which more individuals are born than reproduce. But there is a good argument to be made that most of the time it is a force for stability rather than change, simply trimming off the unwanted or nonadvantageous extremes to keep the entire population as fit as possible. What is more, natural selection perforce operates on entire organisms, rather than on any specific traits that might be perfected. In an unforgiving world it is not of much evolutionary advantage to be the fleetest of foot in your population, or even the smartest of mind, if you are also the most short-sighted or have a low sperm count. The individual is a fantastically complex entity, its development governed by an equally complex genetic and developmental system, and it is hard to see how selection could single out specific attributes to favor except in very special cases, most of them directly tied to reproduction or survival.

One other consideration applies with special force to the human lineage. For the most part natural selection operates slowly, gradually shifting gene frequencies from one generation to the next in response to environmental pressure. Yet research in recent decades has clearly shown that the human lineage evolved during a time of wildly oscillating climatic and environmental settings. The conditions for perfecting adaptation simply were not there. Indeed, the trick of getting by in such conditions lay in avoiding putting all your adaptive eggs in a single basket.

Darwin changed our thinking about evolution by giving us a populational perspective on how variation occurs in nature; without this populational approach to nature, natural selection could never have been discovered. In turn, population thinking also led to the realization that populations evolve, not individuals. And once biologists had come to understand that genes control the traits that we see in nature, a new way of characterizing evolution became available. Before the advent of genetics, most of evolutionary biology was phenomenological and descriptive. But recognizing that a single gene can have alternative forms (called "alleles") allowed for a quantitative approach to studying evolution. Let's say there are two alleles for a particular gene. One allele we will call "A," and the other form is "a." If we have a good sample of a population, we can characterize it as having, say, 50 percent A and 50 percent a. Another population might have 10 percent A and 90 percent a, and so on. Such allelic frequencies have given modern evolutionary biology a highly quantifiable way to characterize or define evolutionary change in terms of changing allelic frequencies in a population. But we must be careful not to be mesmerized by this, because it is not only gene frequencies that change; the genes themselves do—and for a variety of different reasons, as we will shortly see.

Nature has produced some very marked differences in the way that organisms reproduce. (Bacteria are clonal, for example, while we reproduce via sex.) It has also generated many different ways to produce variation. Thus, while most organisms rely on mutations in the germ line (the reproductive cells) to generate variation, bacteria (while also taking advantage of mutation) can also transfer genes asexually. Nonetheless, most organisms on this planet—all the way from bacteria to whales—play pretty much by the same evolutionary rules when it comes to allelic frequencies. And because humans fit fairly comfortably in this spectrum, it is important for us to establish the general ground rules of evolution if we are to understand the biological history of our own species. Which is why we will return to many of these evolutionary lessons throughout this book.

2

• • • •

Species and How to Recognize Them

This is a book about race; and races—useful or not—are subdivisions of species, those basic actors in the grand evolutionary play. Understanding what species are is thus fundamental to understanding what races are, and to do this we need to start at the very beginning.

The goal of all life on this planet is reproduction. This is true whether we are looking at a single-celled microbe or a multicellular organism like ourselves. Even some things that are not traditionally considered life, like viruses and prions (a kind of small protein) have the goal of making more of themselves. It is also the case that all organisms on this planet play by the same biochemical rules. At any given time, a microbial cell has thousands of biochemical interactions going on inside its cell walls; and while humans are much larger than microbes, the main difference between us and them is that the number of biochemical reactions going on inside us at any given moment in time is on the order of trillions. In this interior biochemical symphony, the sheet music in each busy cell is deoxyribonucleic acid (DNA). But like any symphony, there are many factors involved in the final musical product. Our genes require context, direction, and cues for the symphony of the cell to work.

An amazing variety of different things happens at the molecular level, mainly because of the astonishing diversity of the molecules themselves. This diversity expresses itself in the ways in which molecules behave, as much as in how they are structured. The reasons why molecules can behave differently relate to their shapes, sizes, and chemical properties (which themselves are often dictated by shape). DNA is a double-stranded and wonderfully symmetrical molecule whose structure makes it a perfect vehicle for carrying replicable information. Each of the two DNA strands is linear and may be composed of literally millions of subunits called bases. Each of these bases, though, is one of only four basic types. The great diversity of DNA molecules in nature comes from the linear arrangement of these four basic building blocks, known as adenine, thymine, guanine, and cytosine (A, T, G, and C). DNA serves both as the blueprint for all the proteins made by your body and as the means whereby your genes get passed on from generation to generation. For now, we will focus on how DNA replicates, because it is this replication process that is important in how we view variation in evolving populations.

Soon after its double helical structure was deciphered, several other very important aspects of DNA were discovered. One important observation was that the two strands of the double helix are held together by bonds connecting them. When a DNA molecule was made, what is known as a "5-prime" end of one base connected to the "3-prime" end of the next base, and so on. It was also known that the number of As in a DNA molecule was roughly equal to the number of Ts, while the number of Gs equaled the number of Cs. The reasons for this remained a mystery until the double helical structure of DNA was determined. Once it was understood that DNA is a two-stranded molecule, it emerged that every A on one strand matched with a T on the other, and the same was true for every G and C. Further experimentation determined that the two strands of the double helix ran in opposite directions, much like two snakes intertwined, their heads pointing in different directions.

For our purposes, the important thing here is that if you have one strand of the DNA double helix, you have an exact template for the other. So when you are replicating your DNA molecule, just follow the rules: whenever your see an A on the template strand, you place a T on the strand being made from it. If you see a T on the template strand, you place an A on the growing strand; similarly, a G tells you to place a C on the growing strand, and a C on the template tells you to put G on the growing strand. In their 1953 publication describing the double helical nature of

DNA, James Watson and Francis Crick wryly noted: "It has not escaped our notice that the specific pairing we have postulated immediately suggests a possible copying mechanism for the genetic material" (Watson and Crick 1953, 738).

And, by the way, you will also have a mechanism for holding and dispersing the genetic information needed to make the proteins that we will discuss now and later in this book.

The information encoded in DNA is contained both in the "genes" and in other regions of the genome. A gene is simply a circumscribed and finite sequence of DNA that codes for a protein that will have a function in the cells of an organism. Genes are arrayed along the "chromosomes" that can be seen microscopically in the nucleus of each cell; and all the chromosomes in a cell together make up the "genome." Even the simplest bacteria have chromosomes, which are usually single and circular. Eukaryotes like ourselves (see chapter 4) have multiple chromosomes (23 pairs in *Homo sapiens*). The smallest number of chromosomes known in any mammal is found in the Indian muntjac, a small deer-like creature with only three pairs—one pair less than fruit flies have. Obviously, the number of chromosomes has little to do with the complexity of the organism.

Note that, for those sexually reproducing organisms we just discussed, we spoke in terms of *pairs* of chromosomes. This is because we inherit one member of each pair from our mother and one from our father. Most chromosomes are "autosomes," which do not differ between the sexes; but one pair is known as the "sex chromosomes," because of its function in determining the sex of the individual involved. Males have one "X chromosome" and one "Y chromosome" in this pair, while females normally have two X chromosomes. Our 23 pairs of chromosomes each carry several hundred to thousands of genes, the most crowded being chromosome 1, at 2,058 genes. The least populous is the Y chromosome (at 71 genes) if you are a male; whereas if you are a female, your chromosome 21 has the fewest genes, at 234.

Because the replication of DNA is central to the reproduction of organisms, DNA can also help us explain how the great diversity of life evolved. The key here is that the replication of DNA is not always exact. In other words, when DNA is replicated using the base-pairing rules we explained earlier, mistakes are sometimes made. Instead of a T on the template strand dictating that A will be placed on the replicated strand, sometimes a C or a G or even another T sneaks in. If such changes occur in the germ cells (sperm lineage of the father or the egg lineage of the mother), the effects— known as "mutations"—can have far-reaching impacts on the offspring of sexually reproducing organisms like human beings. Microbes like bacteria and archaea reproduce asexually and simply make linear clones of themselves, but mutations are an important part of their biological histories, too.

In clonal organisms, the mutation just needs to happen before the parent cell divides into the two daughter cells, allowing the offspring cells to differ from the parent. But sexually reproducing organisms have two significant characteristics that also affect the nature of the variation among members of a species. Because both the mother and father contribute genetic material to their offspring, the kind of genomic information applying to any single gene can be complex. Take, for example, a hypothetical "Es" gene, for which your mother transfers to you a form we will call "BIG S," and your father transmits a form of the gene we will call "little s." These are the same gene, but are known as different "alleles" of that gene, because they differ slightly in their DNA sequences. Your "genotype" (your alleles) for the Es gene is therefore BIG S/little s, which we can conveniently shorten to S/s. In this case, you are "heterozygous" for this gene; but if your parents had both given you the same allele, you would be "homozygous" (either S/S or s/s). Another complication that may arise and produce new variation in a system like ours is that different parts of chromosomes can cut and paste their DNA with each other, in a process known as "recombination." We will look more closely at this phenomenon when we discuss ancestry.

As offspring give rise to their own offspring, and on down the line, accumulating changes may make descendants look different from their progenitors. Various things may drive this divergence. Natural selection, for example, will look at a new variant or mutation in one of three different ways. It can view the mutation as beneficial, in which case it will favor the propagation of the variant. Natural selection might alternatively view the mutation as deleterious, and will therefore eliminate that variant from the population. Or it might view the mutation as neutral, not caring either way, in which case the variant will hang around. But natural selection is not the only player in the evolutionary game, and many kinds of chance factors may intervene to shift gene frequencies one way or the other or to favor (or not) new variants that have come about through mutation. In either case, if two alleles are present in a population for a specific gene, that population is said to be "polymorphic" for that gene. If events, selective or otherwise, drive one allele or the other toward extinction, the persisting allele is said to be "fixed."

This system may lead to one of several interesting evolutionary outcomes. Natural selection will, very likely, quickly weed out any mutation that is strongly detrimental to its possessors. On the other hand, a beneficial new mutation will have the potential to spread in a population as natural selection favors it; and it might even eventually eliminate the original form from which it mutated. The frequencies of those neutral variants will bounce around, often reaching different population frequencies purely by chance through the "genetic drift" we mentioned in chapter 1. And, very importantly, detrimental, advantageous, or neutral variants resulting from mutation might be amplified in small populations purely through sampling effects. Indeed, small population size may result in some very strange polymorphisms and fixation events.

Now, imagine all this replication, mutation, selection, and drift happening on a grand scale, over long periods of time, and you will get a feeling for the sense of wonder that Darwin was expressing when he referred to "the great Tree of Life." Darwin was clearly mesmerized by the striking tendency of nature to produce diversity through the branching of lineages of organisms; and the subsequent discovery of genetics and its molecular basis has only confirmed his conviction that the major way in which complex organisms diversify is through this branching process. Darwin's one figure in the *Origin* is, indeed, one of the clearest illustrations ever made of this process. It is fundamental to his book, because he saw so vividly how often divergence has led to differentiation and, eventually, to the fantastic variety of body plans we find on this planet.

Since at least the time of Aristotle, humans have understood the importance of species in nature and have wrestled with the problem of how to define them. Our word "species" comes from the Latin word for "kind," which was how natural historians viewed species until the eighteenth century: the "kinds" of organisms that the Creator had placed on Earth. Still, with a century and a half of evolutionary thinking behind us, you might well imagine that by now we would have a finely honed concept of species. Sadly, however, we don't. Modern evolutionary biology is rife with competing species definitions, mainly because different aspects of natural history make different demands. A population biologist who studies species boundaries among mammals will have a different definition of species from a paleontologist who studies dinosaurs. And both will have definitions different from those of an anthropologist who studies fossil hominids. Yet everyone agrees that species are in some sense "real." So how do we go about recognizing species?

Perhaps the first modern, cohesive definition of species came from the great twentieth-century biologist and ornithologist Ernst Mayr, whose concept is cited in every textbook of evolution. Like many natural historians of his time, Mayr recognized the importance of reproductive isolation when looking at species, resulting in his famous compact formulation that "species are groups of actually or potentially interbreeding populations, which are reproductively isolated from other such groups." By dissecting this concept, we can better see where the problems in defining species lie. First, note that Mayr uses the term "groups." What a "group" is in this context remains vague, which is the first major problem. A similar ambiguity attaches to the term "population," although the interbreeding criterion helps. The next difficulty concerns the phrase "actually or potentially." The "actually" part is easy, because it implies observation of the activity concerned. But the word "potentially" is a problem, removing the issue from the realm of dispassionate observation. Come to that, so does "reproductively isolated," for such isolation remains essentially an abstraction, as we discuss later.

Even if you see two individuals engaging in reproductive behaviors, you cannot necessarily know that they belong to the same species, since there is many a slip between cup and lip: for reproductive compatibility to be complete, the offspring (if any) would have to be fully viable and fertile under natural conditions, whatever they might be. To cut a long story short, Mayr's definition is more or less impossible to apply in practice; and this applies to virtually every other definition of species that is based in any way on reproductive criteria. Yet we know that at some level reproductive boundaries are incredibly important in the "packaging" of nature; diversification would have been impossible without such boundaries. And we cannot get around the issue by invoking criteria of other kinds, because if we do this, we find ourselves in the realm of the typology from which Darwin freed us with his population thinking.

So let us return for a moment to the nitty-gritty of collecting data relevant to determining species status. In rare cases, individuals from two established species may indeed mate with each other, but their offspring will either die as embryos, live to adulthood but not be able to reproduce, or reproduce sparingly, maybe with the offspring unviable or unable to reproduce. So even if two suspected species interbreed in nature, and you are lucky enough to observe the behavior—often something exceedingly difficult and time-consuming to do, if not impossible—there is no assurance that the act will result in viable offspring at either of the levels discussed earlier. All of this is just a convoluted way of saying good luck with the operationality of the Mayr concept. It's objective, but in the context of data collection it isn't practical.

Another problem arises because the process of speciation—the splitting of one species into two—occurs over time. We might observe two "populations" at one point in time and discover that they do not interbreed because they are isolated by geographical barriers. These two populations fit the Mayr criterion of reproductive isolation. But some time down the road, they might come into contact again and start to interbreed freely, suggesting that speciation has not taken place. This temporal problem has led our colleague Kevin de Queiroz to suggest that temporal issues might make species concepts a "moving target." In other words, the same species concept might work perfectly well in a certain temporal framework but be entirely inadequate in another. And while in a more perfect world we might hope for a more objective way of looking at species, many think this might be as good as it gets.

Several decades ago, the biologist–philosopher Mike Ghiselin tried to get away from the endless controversy over species definitions by simply characterizing species as "individuals." By this he meant that species are best viewed abstractly, as entities that are launched on their own evolutionary trajectories and are no longer at any risk of integration with other such entities. Since newly differentiated species are very closely related, it might not be surprising to find a bit of hybridization among them, especially at early stages in their separate histories; but if such interbreeding is rare, or unsuccessful enough not to lead to eventual reintegration, it will remain irrelevant to species status. For example, our big-brained close relative *Homo neanderthalensis* was the resident hominid in Europe and western Asia some forty thousand years ago, the point at which our own species *Homo sapiens*, emerging from Africa, invaded these regions for the first time. Early on, there may have been occasional interbreeding between the two cousins—which is why various agencies today offer, for a fee, to determine what percentage of "Neanderthal genes" you yourself possess (see chapter 9). Nonetheless, the two hominids remained morphologically, culturally, and behaviorally distinct, right up to the time when the Neanderthals finally disappeared. Any interbreeding there may have been—and some believe there are alternative explanations for those Neanderthal genes of yours—was evidently immaterial to the long-term fate or evolutionary trajectory of either party.

Ghiselin's abstract solution seems to be preferable in cases such as this, but it is nonetheless sometimes necessary to have a method that can recognize species on the ground. Conservation biology is a case in point. This discipline attempts to bring all arms of biology to bear upon the task of conserving the biota and managing fragile areas and endangered or threatened species. It is often described as a "crisis discipline," with similar import to cancer or HIV biology. In any crisis discipline time is the enemy, and as much of modern science as possible is applied to the problem at hand. For those charged with protecting endangered species, the operationality of the species concept used is obviously at the forefront; and no matter how philosophically sound one's criterion for species designation, it had better be operational.

One way of achieving this is to come up with a sound species concept, to determine what the major practical implications of that concept are, and to specify the resulting criteria for species recognition. For example, if we take Mayr's "biological species concept," we can predict that any two populations that conform to it will be genetically differentiated in some way. But what do we mean by "differentiation"? One might claim that two populations that have speciated should have a specifiable degree of differentiation between them. But here subjectivity intrudes. One researcher might claim, for example, that a 2 percent difference in genetic sequence is a good indicator of species status. In other words, if the genomes of individuals of two populations are one hundred bases long, an average of two bases separates the two populations, and there are fewer than two base differences within each population, then we have a reason to conclude that the two populations represent two different species. But it is entirely possible that another researcher will come along and demand a 3 percent difference, or that yet another will demand 10 percent. This is not a notably objective procedure, and clearly a well-defined either/or criterion is needed.

So let us think again about the hypothetical populations we discussed earlier. If it turns out that at least one of the two bases is fixed in the two populations, and is as distinct as shown in figure 2.1, then the ability it

```
                                      1
                              9999999990
                              1234567890
        Population 1
        Individual 1  . . . gcatcgtcaa
        Individual 2  . . . gcatcgtcaa
        Individual 3  . . . gcatcgtcaa
        Individual 4  . . . gcatcgtcaa
        Individual 5  . . . gcatcgtcaa

        Population 2
        Individual 1  . . . gcatcgacta
        Individual 2  . . . gcatcgacta
        Individual 3  . . . gcatcgacta
        Individual 4  . . . gcatcgacta
        Individual 5  . . . gcatcgacta
```

Figure 2.1 The last ten bases of the genomes of the individuals in the hypothetical populations discussed in the text. The first ninety bases of the genome are identical in all individuals from both populations, and so they are represented by the dots. The numbers above the sequences refer to the positions of the bases in the one hundred base-pair genome. The diagnostic base is the next to last one and is in bold.

confers to diagnose the two populations might be regarded as sufficient for recognizing that the two populations have speciated. And indeed, diagnosis using morphological attributes of organisms is how taxonomy has been done since Carolus Linnaeus invented the science in the mid-eighteenth century.

But the diagnostic approach also has its shortcomings. First, the fewer individuals one uses to create the diagnostic, the weaker the conclusion will be. Imagine examining two populations with five individuals in each population and finding a diagnostic that is both fixed and different between the two populations. It is very likely in this case that if you go on a collecting trip a week later and collect two more individuals from each population, the diagnosis will fall apart, because you have collected some new individuals that do not have the diagnostic (figure 2.2). Additionally, if you examine more populations, the diagnostic has a higher likelihood

```
                                        1
                               9999999990
                               1234567890
Population 1
Individual 1   . . .  gcatcgtcaa
Individual 2   . . .  gcatcgtcaa
Individual 3   . . .  gcatcgtcaa
Individual 4   . . .  gcatcgtcaa
Individual 5   . . .  gcatcgtcaa
Individual 6   . . .  gcatcgacta
Individual 7   . . .  gcatcgacaa


Population 2
Individual 1   . . .  gcatcgacta
Individual 2   . . .  gcatcgacta
Individual 3   . . .  gcatcgacta
Individual 4   . . .  gcatcgacta
Individual 5   . . .  gcatcgacta
Individual 6   . . .  gcatcgacaa
Individual 7   . . .  gcatcgacta
```

Figure 2.2 Increasing sample size has the potential to destroy diagnosis.

```
                                      1
                               9999999990
                               1234567890
Population 1
Individual 1  .  .  .  gcatcgtcaa
Individual 2  .  .  .  gcatcgtcaa
Individual 3  .  .  .  gcatcgtcaa
Individual 4  .  .  .  gcatcgtcaa
Individual 5  .  .  .  gcatcgtcaa

Population 2
Individual 1  .  .  .  gcatcgacta
Individual 2  .  .  .  gcatcgacta
Individual 3  .  .  .  gcatcgacta
Individual 4  .  .  .  gcatcgacta
Individual 5  .  .  .  gcatcgacta

Population 3
Individual 1  .  .  .  gcatcgacaa
Individual 2  .  .  .  gcatcgacaa
Individual 3  .  .  .  gcatcgacta
Individual 4  .  .  .  gcatcgacta
Individual 5  .  .  .  gcatcgacta
```

Figure 2.3 Increasing the number of populations has the potential to destroy diagnosis.

of being destroyed (figure 2.3). And these caveats more than likely apply to any criterion one might derive from a species concept.

In this example, the diagnosis was upset by adding more data. Does this mean that the initial diagnosis was invalid when it was made? Actually, no. The inference made at the time the diagnosis was first made was perfectly logical and valid, because all taxonomic conclusions are hypotheses. And, like all scientific hypotheses, they are susceptible to being falsified by more or better data. This example points to an important aspect of systematics and taxonomic science: namely, they are revisionary and subject to constant review. If we had collected all the individuals of the populations concerned, and no other populations were ever found to exist, then revision would not be possible, and the original diagnosis would have to stand. But as long as there are more of the same organisms out there, the possibility of a revised diagnosis—or of additional corroboration of the original one—is always there.

Species, then, may be slippery things; and the packaging of nature, while clearly real in a fundamental sense, is much less tidy than the orderly minds of systematists might like. Nonetheless, we suggest that with the use of appropriate methods, most species, at least in the living world, can be detected with fair confidence. Indeed, it often turns out that diagnosis is a reasonably good proxy for the biological species concept, especially when bolstered by as much other evidence as possible, including that of behavior, ecology, and geographical distributions.

But what about the further subdivision of a species into subspecies, or races? Here we are in very tricky territory indeed, because we can see no objective way to address those hypotheses of subspecific differentiation. At this level the reproductive criterion does not apply, and any other criterion one might attach to the delineation of subspecies or races will be entirely subjective. One distinguished biologist once declared that "subspecies may be recognized if they are useful to the taxonomist," pretty much conceding that they have no objective existence. Accordingly, in most contexts the terms "subspecies" and "race" are used mostly when a researcher is unsure of the status of a population. In other words, they are generally used in taxonomy as highly provisional hypotheses of species existence. And logically, if one can reject the hypothesis that there is a species boundary in a system, then the hypothesis is discarded and one can move on.

Where does this leave us in the case of *Homo sapiens*? When Linnaeus named and diagnosed our species back in the eighteenth century, he did it with the enigmatic comment *nosce te ipsum*, "know thyself." And though this would normally be rather unhelpful in taxonomic terms, it is no impediment in practice to recognizing the boundaries of our species, because we are the lone surviving twig of what was once a very luxuriantly branching hominid tree. On the outer side, we have no close living relatives who might make a claim on our identity; and on the inner one, it is abundantly clear that we fulfill Mayr's requirement of interbreeding: something we do liberally whenever populations of different geographic origins meet. Linnaeus himself recognized the geographic divisions *Americanus,*

Europaeus, Asiaticus, and *Afer* (African) within *Homo sapiens;* but then, he recognized the entirely imaginary *Monstrosus* and *Ferus* as well. So while there is no doubt that there has been some differentiation within our species, as we will see in later chapters, this differentiation is extremely recent, entirely superficial, impossible to reliably diagnose, and has no bearing whatsoever on our reproductive status.

3

• • • •

Phylogenetic Trees

Both the sole illustration in the *Origin* and the famous "I think" tree were drawn from Darwin's imagination. In other words, neither diagram represented relationships within a group of real organisms. Nor did what we believe to be the very earliest tree diagram, drawn by Jean-Baptiste Lamarck and published in his *Philosophie Zoologique* of 1809 to represent his view of how life had evolved on the planet. The most famous early tree that represented organisms in the real world was the artistically gifted Ernst Haeckel's "Pedigree of Man" (figure 3.1), in which he took the phylogenetic tree to its extreme metaphorical manifestation by drawing an actual tree with a trunk, limbs, and bark and placing the organisms on the tips of the branches. But again, no formal analysis was presented in support of this tree. And so it went, until around the middle of the twentieth century.

Close examination of the trees drawn by Lamarck and Haeckel reveals that both scientists liked to put living species at the branching nodes of the trees they drew. The strong implication here is that one group had evolved into another until humans achieved the pinnacle of evolution at the very top of the tree. We now know that this approach to constructing phylogenetic trees is the wrong way of going about representing actual evolutionary relationships. In fact, the concept of a tall, slender tree is inherently mistaken; a much better analogy is a luxuriantly branching bush without a single central stem. Living taxa lie at the very tips of the

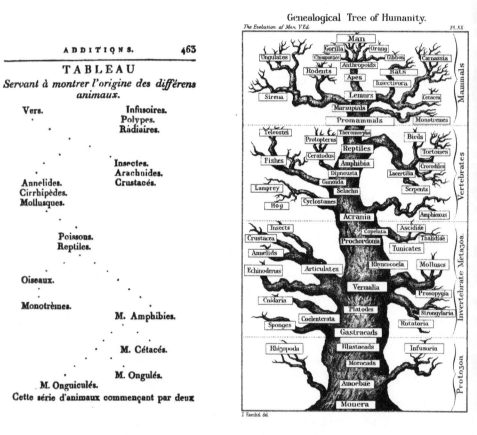

Figure 3.1 Two early versions of evolutionary trees. Lamarck's tree from 1809 (*left*) and Haeckel's tree from his *Evolutionary History of Man*, published in 1874 (*right*).

branches, while the branching nodes below them are occupied by ancestral forms. For the most part those ancestors will be hypothetical, their existence inferred from that of their descendants. It is possible that the ancestor will be known from fossils preserved in the fossil record; but while plenty of fossils are known that are related to living forms—and to each other—it turns out that most fossils don't satisfy the stringent anatomical requirements that would indicate they are directly ancestral to specific living forms. Instead, fossils normally branch off into their own lineages. Of course, this doesn't mean that fossils are useless in phylogenetic analysis. They are basic to our understanding of the events of the evolutionary past, and they can clarify all sorts of evolutionary questions.

We entered our careers in biology and systematics about ten years apart, during the period of great tumult in systematics that occurred in the 1960s and 1970s; and both of us were influenced greatly by this revolution. To cut a long story short, before this time evolutionary trees were generally drawn by experts who rarely felt it necessary to justify the results of their intuitive judgments. Researchers who worked in systematics were expected to know their groups well enough to conjure up the relationships from their knowledge of evolution and of the species they were studying. In retrospect, this approach gained the name of "evolutionary taxonomy."

During the 1960s three major schools of systematics emerged that energetically and efficiently rejected evolutionary systematics, realizing that an objective and repeatable algorithm for assessing evolutionary relationships was long overdue. These three schools, known rather forbiddingly as phenetics, maximum parsimony, and maximum likelihood, then proved Sayre's law ("In any dispute the intensity of feeling is inversely proportional to the value of the issues at stake. That is why academic politics are so bitter.") by proceeding to battle with one another ferociously. It was an intellectually invigorating process, but not necessarily a pretty one, and bad feelings still tend to linger. Not that the three camps disliked each other for trivial reasons, because deep issues of philosophy underpinned their differences. But the passage of time has provided perspective, and today most systematists have begun to ignore the controversy and to tend instead toward pluralism in the way in which they construct phylogenetic trees—even though the assumptions and philosophical approaches of the three camps remain at odds with one another.

Before we go any further in our discussion of phylogenetic trees, we need to introduce some terminology. First, modern trees are constructed using data that usually come in the form of what are called "characters." Characters are biological features of organisms and may be expressed in a variety of ways. For instance, a character might be the presence or absence of a structure in the taxa being analyzed. Or, if DNA sequences are at issue, a character might be the base that is present at a specific position in the DNA sequences involved. Characters are the currency of phylogenetic analysis. Phylogenetic trees contain branches, and the taxa in the analysis lie at the tips of those branches, which are called "terminals." The points below them, where the branching occurs, are called "nodes," and they equate to the ancestors of the taxa in the tree. Thus, if lions and tigers lie next to each other in a tree of big cats, the node that connects them represents the common ancestor of lions and tigers, which, as each other's closest relatives, are known as "sister taxa." The very bottom of the tree is called its "base," and the lowest node is called the "root."

Frustrated by the lack of objectivity in evolutionary taxonomy, in the late 1950s and early 1960s the quantitatively minded biologists Robert Sokal and Peter Sneath invented an approach to systematics that they called "numerical taxonomy." This used algorithmic, quantitative approaches to constructing phylogenetic trees. As Sokal and Sneath saw it, the advantage was that numerical taxonomic approaches were objective and repeatable and could be automated (and thus be highly operational). A researcher interested in a specific group would start to collect information about the specimens being studied. If he or she were working on fish, for example, such information could range from the number of scales on a specific structure, to whether or not a fin was in a particular place, to the shape of teeth, to the morphology of an internal organ. Because the structure of DNA had been discovered only a few years earlier, Sokal and Sneath had little to say about genes; but in fact DNA sequences are also perfectly amenable to this approach.

Once the information was collected, an algorithm could be used to generate metrics representing the degrees of overall similarity among the taxa of interest. These were then used to construct a diagram representing those degrees of similarity. The tree-construction step was relatively simple and mathematically easy to perform. The organisms with the greatest overall similarity were placed together as each other's closest relatives and connected by a node. The similarity metrics of the rest of the data set then determined where a third taxon was attached to the tree, and so on. The big assumption in this process was that overall similarity given a particular character set equated with phylogenetic affinity (common ancestry); the big drawback was that this might not always be true—indeed, it often was not. In terms of ancestry a lungfish is more closely related to a cow than it is to a salmon; but on most data sets, numerical taxonomy would place the two aquatic forms together.

Soon after the end of World War II, a German entomologist named Willi Hennig began thinking about the drawbacks of evolutionary taxonomy and came up with his own approach to building trees. He published his ideas in a book entitled *Phylogenetic Systematics* in 1950, but it was not until 1966 that his ideas were translated into English and came before a worldwide audience. Thereafter, they were taken up by an enthusiastic band of systematists in the United States and Great Britain.

The core of Hennig's system lay in the realization that divergence via common ancestry would leave more closely related taxa sharing some of their characters to the exclusion of other taxa. For instance, mammals have hair, which other organisms don't have, because they have inherited it from the hairy ancestral mammal. Hair is therefore not only a shared attribute of mammals, but a "derived" one, because it differs from the condition present in the common ancestor's own predecessors. In Hennig's terminology, as a departure from the "plesiomorphic" (primitive) general vertebrate condition, hair is an "apomorphy"; and it is a mammal "synapomorphy," because it is universally present among mammals (except for a handful that have secondarily lost it) and is thus a character that binds all mammals together.

As an ancestral retention, hair is also "homologous" among mammals. But not all similarities are homologous. Consider the wings of birds, bats, and insects. Grouping these very different creatures together because all share the ability to fly, to the exclusion of other mammals (the group that bats are in), other arthropods (the group that insects are in), and other archosaurs (the group birds are in), would allow us to hypothesize that wings evolved once, just as hair did in mammals. But since we have abundant evidence to demonstrate that this is not the case, it is by far the simplest hypothesis to propose instead that wings evolved three times, independently, in the ancestors of the bird, bat, and insect lineages. To use Hennig's term, wings in this instance are a "homoplasy." A more commonly used word would be "convergence," or, as Darwin would have said, "analogy."

As it turns out, both Hennig's ideas about phylogeny and the approach that was eventually adopted to implement his thinking are objective, repeatable, and operational. In understanding the lines of attack that have

Figure 3.2 Three possible hypotheses concerning the relationships of lions and tigers and bears. Left shows tiger with lion excluding bear, middle shows lion with bear excluding tiger, and right shows tiger with bear excluding lion.

been taken to implement Hennig's ideas, the first step is to realize that trees may take many different forms (topologies). If you have three taxa (a lion, a tiger, and a bear, say), then there are three possible topologies that need to be examined to figure out which one best fits the data and thus presumably best reflects the actual evolutionary sequence of events: tiger with lion excluding bear, tiger with bear excluding lion, and lion with bear excluding tiger (figure 3.2). As the number of taxa increases, the number of possible trees rises rapidly. By the time you have fifteen taxa in an analysis, there are trillions of trees to evaluate.

What is more, we need a way to give direction to the tree, because without knowing which of the three species is most ancestral (that is, has characters closest to the ancestral state) we can't determine which of the three possible trees is best. We can arbitrarily decide which of the three taxa is the most ancestral, or we can be more objective and include in our analysis a fourth taxon that lies outside the lion, tiger, and bear group. For example, a monkey, a sea lion, or some other mammal. While a sea lion might at first glance seem best to root the tree, because it is a carnivore as well, it is not ideal, because the best roots are "outgroups" (closely related, but not actually a member of the group under study), and the sea lion actually belongs to the carnivore "ingroup." So another species is needed. A monkey could be used, but it is rather distant from the three ingroup carnivores. And since it turns out that pangolins are much more closely related to carnivores, let's use them as our outgroup

	m	1	2	3	4	5	6	7	8	9	0	1	2
Lion	+	ATG	AAT	TAT	ACA	AGT	TAT	ATC	TTA	GTT	TTT	CAG	CTC
Tiger	+	ATG	AAT	TAT	ACA	AGC	TAT	ATC	TTA	GCT	TTC	CAG	CTT
Bear	−	ATG	AAT	TAC	ACA	AGT	TTT	ATT	TTC	GCT	TTT	CAG	CTT
Pango	−	ATG	AAT	TAC	ACA	AGT	TTT	ATT	TTC	GCT	TTT	CAG	CTT

	m	1	2	3	4	5	6	7	8	9	0	1	2
Lion	+	M	N	Y	T	S	F	I	F	A	F	Q	L
Tiger	+	M	N	Y	T	S	F	I	F	A	F	Q	L
Bear	−	M	N	Y	T	S	Y	I	L	A	F	Q	L
Pango	−	M	N	Y	T	S	Y	I	L	V	F	Q	L

Figure 3.3 Data matrix for lions and tigers and bears (with pangolin = pango as outgroup). The tops shows DNA sequence with a morphological character (m). The bottom shows the translated amino acid sequences for the DNA sequences on the top. The gene used is the interferon gamma gene, and only the first twelve codons are shown. There are 167 codons in this protein in humans. The morphological character is "purrs" where a "+" indicates yes and a "−" indicates no. The numbers above the sequences refer to the codon position. Red and blue in a position indicates that position implies the tigers and lions belong together. Green indicates a position that is variable but has no information as to how the ingroups are arranged. For color figure, see plate 1.

Figure 3.3 shows real data from lions, tigers, and bears, with a pangolin included as an outgroup. DNA information comes from the interferon gamma gene and protein, and a morphological character (m: on whether the organisms "purr" or not) is also included. Note that there are five characters (one morphological and four DNA) that support lion with tiger, zero characters that support bear with lion, and zero characters that support tiger with bear. For the protein translation of the DNA sequences in the figure, there are three characters (one morphological and two protein) that support lion with tiger, zero characters that support bear with lion, and zero characters that support tiger with bear.

Using these data, the bottom line is that the tree showing tigers together with lions is the one best supported by the data (figure 3.4), taking the fewest evolutionary steps to explain and fitting the criterion of "maximum parsimony." This result is hardly surprising, as we know that lions and tigers are both in the family Felidae, while bears are in the separate family Ursidae. In other words, we knew beforehand that cats purr and dogs bark. Rooting with a pangolin was wise, because they are a close outgroup to lions, tigers, and bears but are in the separate family Pholidota.

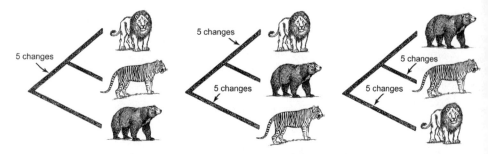

Figure 3.4 The three phylogenetic trees in figure 3.2 showing where the four DNA sequence changes and one morphological change ("5 changes") need to occur on the tree to explain the data. The tree on the left takes five steps, the one in the middle takes ten steps, and the one on the right takes ten steps, making the one on the left the "most parsimonious." For the protein sequence, one simply needs to substitute "3 changes" for "5 changes" in the figure.

At this point the astute reader will be asking one big question, namely, why use the parsimony criterion to determine the best tree? After all, evolution doesn't necessarily have to be parsimonious. We have already argued that evolution can be quite tangled, and this untidiness may well include complications due to convergence and homoplasy. So, while the criterion of parsimony seems to be a good starting point, we need to keep in mind that it does incorporate an arbitrary assumption about evolutionary history.

Recognizing this, the biologists Luca Cavalli-Sforza and Anthony W. F. Edwards formulated the third camp of modern systematics in the later 1960s. They suggested that the events involved in the evolutionary process can best be estimated using statistical likelihood, or the probability that events will occur. Specifically, they developed the method for molecular events in which the probability of changing one base into another is designated by a model. The content of DNA sequences and the chemistry of the bases makes it reasonable to put probabilities on such base-change events. For instance, we know from the chemistry of DNA that a mutation from an A to a G has a higher probability than from an A to a T. Other aspects of the probability of DNA sequence change are also used, and models can become quite complicated when applying them to phylogenetics. Currently, there are more than 200 models on offer for maximum-likelihood analysis.

The maximum-likelihood approach computes the likelihood of the tree given a specific model and a particular data set. When a maximum-likelihood analysis is performed, the likelihood of the data is computed for all possible trees using a given model, and at the end of the day one can pick the tree that has the highest likelihood. This approach is preferred by researchers who feel that parsimony is too restrictive an assumption, while parsimony advocates prefer their method because they feel that adding more parameters in a model adds more assumptions and constitutes a slippery slope away from strictly scientific hypothesis testing.

In the 1990s another statistical approach was developed. Currently used by a lot of researchers not only in phylogenetics but also in population genetics and species delimitation studies, it relies on Bayesian statistics. The Bayesian approach usually incorporates a likelihood model, a major difference being that Bayesian analyses use prior probabilities of the existence of each tree. The tricky thing about Bayesian analysis is that it doesn't search for the best tree. Instead, the analysis attempts to create a probability distribution of the possible trees and then assesses the probability of the branching events from this probability distribution. We can, for example, compute a probability distribution of phylogenetic trees for mammals based on DNA sequences. If we want to know the probability that lion and tiger are each other's closest relative, then we simply scan the distribution and count the number of times lion and tiger are found together in the trees in the distribution. Because of their ability to incorporate all prior knowledge about evolutionary events, Bayesian approaches are being applied more and more in modern evolutionary analysis.

Although these various methods work in different ways, we are happy to report that highly divergent results using different methods on the same data set are rare. For the most part, trees generated by the different methods agree in basic ways, which suggests that there are consistencies in the evolutionary process and that all the methods are pretty good at extracting the historical information encoded in the data.

Once we have a phylogenetic tree for a group of organisms, a lot of very important evolutionary questions can be addressed. We have already discussed the one that concerns reconstructing ancestors. Since ancestors occur at the nodes of a tree, reconstructing what an ancestor was like is a simple process of deciphering what character changes have occurred at those nodes. Figure 3.4 shows a simple example of this method using parsimony, though likelihood and Bayesian methods are equally applicable. Another important use of phylogenetic trees is in determining whether two different taxa can be included in the same higher group. If an analysis includes several closely related species from a few genera, the branching patterns in a phylogenetic tree can tell us which species belong in which genera. Taxonomic groups are recognized by ancestry, meaning that all members of a given group must be more closely mutually related than they are to members of any other taxon. This brings a final term into our lexicon: "monophyly," denoting descent from a single common ancestor.

Importantly, within the context of this book, a phylogenetic tree of organisms from discrete geographic regions can also be used to address not only evolutionary history, but also questions about biogeography and how organisms have moved around on our planet. This has special resonance in our attempts to understand how our own recently evolved species contrived to exit from its natal continent of Africa and, in an amazingly short time, take over the entire habitable Earth.

Finally, a phylogenetic tree can be used to estimate the time at which two lineages diverged. Two methods are used here. The first involves establishing the minimum latest occurrence of a group from fossil evidence and then inferring the age of the group and subsequent divergences from that date. Another approach is to use the "molecular clock" method, based on the assumption that DNA sequences change in a regular, clocklike manner. With this assumption, if one knows the age of divergence of two taxa based on a fossil date, one can use this age to calibrate the number of changes that typically occur in a specific DNA sequence over a certain period. This calibration can then be used to estimate divergence times at other nodes that lack fossil dates.

This discussion of the uses of phylogenetic trees in the arsenal of evolutionary analysis is by no means exhaustive; but the uses outlined here are the basic ones that we will depend on throughout this book. The history of our species on this planet has been deciphered largely thanks to such approaches; and much of our knowledge about the history of other organisms is also contained in phylogenetic trees. But, as always, we need to be careful. Sometimes a tree is not the most appropriate metaphor for reconstructing evolutionary histories; and while tree building is an entirely appropriate approach to establishing the context within which human beings evolved, it should not be applied unquestioningly to the extraordinary events that were involved in the generation of current human diversity.

4

• • • •

The Name Game

MODERN ZOOLOGICAL NOMENCLATURE
AND THE RULES OF NAMING THINGS

Humans have developed naming things into an art. Of course, we are not the only animals to have names for objects and other organisms we encounter. For instance, prairie dogs and some monkeys can communicate the kind of predator they are confronting through emitting specific sounds. But no other species names things in the way we do. We have named natural objects at least since our precursors invented language, probably around one hundred thousand years ago; and the habit of classifying them via folk taxonomy—and thus ultimately via taxonomic principles—certainly has its roots in language itself. Naming things is a very deeply embedded component of our cultural and evolutionary development.

The world is a very untidy place, and it must have seemed even untidier to our ancestors who were just taking up language and trying to understand their surroundings in the new way that language permitted. Nonetheless, the journalist Carol Yoon has suggested that our brains are hardwired to name things. She describes humans as living in an "umwelt" (self-centered world), which forces us to create taxonomies to make sense of our environment. This umwelt must have driven our ancestors to give names to objects around them rather quickly, since some of the animal and plant life they lived among could be extremely dangerous.

It is almost impossible for us to imagine what it was like for the earliest linguistic *Homo sapiens* to go through the process of organizing the world around them into a vocabulary of names that they could then juggle and conjure with; but this was certainly the most fateful cognitive shift in the history of any organism. By the time of Aristotle, whom we have already mentioned in the context of typological thinking, this initial process was far in the past; but typology had almost certainly always been at the heart of the process of naming objects, particularly the naming of organisms. Aristotle named more than five hundred animals that were familiar to him and his contemporary ancient Greeks, and he was in fact very utilitarian in how he named them. His desire was simply to be precise for his readers, so that they would know what he was talking about in his broader philosophical disquisitions.

In science, names are given to species and other natural units according to a set of rules agreed over the last three centuries or so. The names themselves are arbitrary, although they are usually thematic in some way. Historically, taxonomic groups have often been named for features of the organisms concerned, as in the case of our own order Mammalia, the name of which incorporates one of the key diagnostics for this higher group: the mammary gland through which newborns are nourished (Linnaeus apparently changed the name of the group to Mammalia from Quadrupedia as a political statement about wet-nursing). Taxonomic names can also signify the specific place or the broader geographical region where the organism is found, as in the case of *Ulex europaeus*, the European gorse plant. Or the name a taxonomist gives an organism might honor its discoverer or another admired person. Barack Obama did pretty well in having nine species named after him during his eight years as president of the United States, though it is probably the nineteenth-century German naturalist Alexander von Humboldt who holds the record, with more than one hundred species bearing his surname. But taxonomy consists of a lot more than simply awarding names to organisms, and indeed it has a pretty complex philosophical and historical context.

Carolus Linnaeus is said to have been quite enamored with himself. And not without reason, because during the eighteenth century he created something of enormous scientific value that is still in routine use today, almost three hundred years later. Because of the vast knowledge of the diversity of nature that he accumulated during his lifetime, Linnaeus was able to come up with a consistent way of naming living things: a system we refer to today as "binominal nomenclature," because it awards each species a combination of two names. The first name is that of the genus the species belongs to, and the second is that of the species itself. The genus name is unique, so that although the species name can be used repeatedly (gorse is not the only "*europaeus*" around), the combination of names is unique for every species. Our binominal name, for example, is *Homo sapiens*, and the two names together designate us and nobody else.

Linnaeus also clearly recognized that variation occurred within species, and he occasionally named varieties within the species he designated. These variants have subsequently come to be known as "subspecies," and they are designated by a "trinomen," in which a third name is added to the species binomen. Thus, when it was fashionable to place the distinctive and now-extinct Neanderthals in the same species as us (an insult to the Neanderthals, we believe), it would be common to see living humans assigned to the subspecies *Homo sapiens sapiens*, to distinguish them from *Homo sapiens neanderthalensis*.

Linnaeus was equally aware of the nested quality of the way in which nature is organized. And he devised a system of higher categories to express it, whereby genera are grouped together into families, families into orders, orders into classes, and classes into phyla. Since Linnaeus's time, other higher categories have been added to this "Linnaean hierarchy," for example by inserting "superfamilies" between families and orders, and by grouping phyla into higher categories called "domains" and "superdomains." Additional ranks of this kind continue to be added to the roster, as researchers feel the need to express in their classifications their expanding appreciation of the diversity of nature.

Let's look at this system, using our own species to guide us through the many names that apply to us in the Linnaean hierarchy. First and foremost, despite the billions of bacterial cells that live in and on us, we humans are "eukaryotes." Eukaryota is one of the three great domains of life: Bacteria, Eukaryota, and Archaea. Eukaryotes differ from members of the other domains in having their genetic material encased in what is called a nuclear membrane. And they have turned out to be more closely related to the singled-celled Archaea than to Bacteria, despite the fact the latter two domains share the lack of a nuclear membrane.

TABLE 4.1
Mayr's Definitions of Classification Terminology

Systematics	The scientific study of the kinds and diversity of organisms and of any and all relationships among them
Taxonomy	The study of how classification works, including its basis, principles, procedures, and rules
Zoological (or botanical) classification	The ordering of animals into groups (or sets) on the basis of their relationships
Zoological (or botanical) nomenclature	The application of distinctive names to each of the groups recognized in any given zoological (or botanical) classification

This observation introduces the need to explain the difference between naming and ranking a group and figuring out the relationship of that group to others. Once again, Ernst Mayr played an important role in clarifying this. He defined the important terms "systematics" and "taxonomy," but added two other important terms, as shown in table 4.1.

What we have done so far here is to call upon three of Mayr's procedures. We have been able to systematize the three domains of life, because we have acquired a plausible and data-supported scheme of relationship. We have not used a phylogeny to do this, but in most cases phylogenies are used to accomplish systematic analysis. We have also dabbled in classification. And finally, we have snuck in a little nomenclature by discussing the proper names for the three domains. The only procedure we have not yet mentioned is taxonomy, but we will introduce this subject in the next few paragraphs as we detail where we fall in the classification of the great Tree of Life, something that is summarized in table 4.2.

Domains are divided into kingdoms. Eukaryota has four kingdoms within it: Plantae, Animalia, Fungi, and Protista. Of these four kingdoms, three are monophyletic (all members descended from a single common ancestor) and pose no basic problems of classification and systematics. The exception is Protista, which is not monophyletic. Fortunately, protist systematics is not our problem here, because as human beings we belong to the kingdom Animalia (often also referred to as Metazoa). Depending on whom you consult, there are between thirty-two and thirty-nine animal phyla. These phyla can be organized into three larger categories that include our own, the superphylum Deuterostomia. This also contains the

TABLE 4.2
Hierarchical Classification of Our Species

Kingdom	Animalia
Subkingdom	Bilateria
Infrakingdom	Deuterostomia
Phylum	Chordata
Subphylum	Vertebrata
Infraphylum	Gnathostomata
Superclass	Tetrapoda
Class	Mammalia
Subclass	Theria
Infraclass	Eutheria
Order	Primates
Suborder	Haplorrhini
Infraorder	Simiiformes
Superfamily	Hominoidea
Family	Hominidae
Subfamily	Homininae
Genus	*Homo*
Species	*sapiens*
Subspecies	*sapiens*

Note: The categories in bold are those most commonly used to classify organisms. These levels can be remembered with the mnemonic "King Philip came over from great Spain." All other levels of classification are used by specialists who are interested in further subdividing the hierarchical classification.

phyla Xenacoelomorpha (strange animals with no brain, but a rudimentary digestive system) and Echinodermata (starfish, sea urchins, and the like).

Within the deuterostomes we belong to the phylum Chordata, which has two infraphyla, Urochordata and our own Craniata, which is in turn divided into two subphyla, Hyperotreti (hagfishes) and Vertebrata (our subphylum). Within Vertebrata there are many groups, of which most are now extinct. The only two groups (superclasses) that are alive today are Gnathostomata (the jawed vertebrates among which we belong) and Hyperoartia (lampreys). Within the gnathostomes, we belong to a group called Amniota, whose members have an amniotic sac around their developing embryos. Amniota is in turn divided into a large number of classes, of which ours is Mammalia.

Still with us? Following the expanded modern Linnaean hierarchy down to the level of our (or any) species is an exhausting affair, and alas we are not done yet. The class Mammalia is subdivided into three subclasses, of which ours is Eutheria, the placental mammals. This subclass contains around twenty orders, again depending on the source. Our order is Primates, within which we belong to the suborder Haplorhini, which also contains the tiny and enigmatic tarsier of southeastern Asia (Tarsiiformes), the monkeys of both the Old and New Worlds (Cercopithecoidea and Platyrrhini, respectively), and our superfamily Hominoidea, which besides us embraces the Southeast Asian gibbons and our closest living relatives—the great apes. Many authorities group living humans along with the orangutans (genus *Pongo*), the gorillas (genus *Gorilla*), and the chimpanzees and bonobos (both genus *Pan*) in the family Hominidae, and separate human beings out in their own subfamily Homininae. There are several genera of hominins, but all of them except our own *Homo* are extinct. Similarly, our own *Homo sapiens* ("man the wise"—blame Linnaeus) is only one of many documented species of *Homo*, and it is the first in a very long time to have the entire world to itself.

The biota on our planet is incredibly rich and diverse—especially if you factor in all those weird and wonderful groups that are now extinct—so it is hardly surprising that classifying it all is such a complicated business and that our understanding of natural diversity is still very much a work in progress. But while we need to recognize and live with the fact that many systematic issues will always be matters of contention, there is a real necessity for stability and objectivity when it comes to the matter of names. We must at least agree on what we are arguing about. So names in the Linnaean hierarchy live and die by the rules established by nomenclature boards that were formed by international agreement to oversee the naming of organisms on this planet. For eukaryotes, there are two nomenclatural systems: the International Code of Nomenclature for algae, fungi and plants; and the International Code of Zoological Nomenclature (ICZN) for animals. It has actually turned out to be a little strange that plants, fungi, and algae are lumped together here, because they actually represent three different groups, one of which (fungi) is actually more closely related to animals; but in practice the system works pretty well. The codes were established to minimize confusion in naming new species and the problems caused by renaming or revising a group based on new data; and the rules themselves are quite strict: not adhering to either code can result in the banning of any name that was improperly proposed.

A quick look at how our species got its zoological name and some subsequent controversy will illuminate how the process works and how deeply the naming of anything depends on history. In many modern sciences, almost anything written over a couple of years ago is rapidly forgotten; but in taxonomy history is omnipresent, and Linnaeus remains among the most regularly cited figures in the scientific literature. In 1758 he published the definitive tenth edition of his great treatise *Systema Naturae*, in which he named nearly ten thousand species of animals and plants. In the section titled *Regnum Animalia* (Animal Kingdom), he applied the species name *Homo sapiens* to living humans. Curiously, while Linnaeus provided a thumbnail diagnosis based on anatomy for all the other species he named, in our case he just made that enigmatic remark we've already mentioned, "know thyself." Clearly, Linnaeus was intimate with his subject and thought no further discussion was necessary (had he not, he would probably have had to invent psychoanalysis). And at the time it wasn't: he was making the rules as he went along, and as the founder of modern taxonomy his names automatically stick unless it is not possible to determine from his descriptions exactly what species he was writing about (clearly not the case with *Homo sapiens*). Nowadays, of course, his cryptic comment about *Homo sapiens* wouldn't fly; and what's more, according to the rules of nomenclature as later developed, every species proposed needs what is known as a "holotype" or "type specimen," an example that is designated and archived in a repository so that other putative members of the species can be compared to it. But that is now, and then was then, and Linnaeus simply didn't indicate such a specimen.

Maybe in our case this shouldn't matter much. After all, Linnaeus was entirely correct in assuming we should know who we are (and, even more importantly, who we want to reproduce with). But complications have arisen anyway. In the interests of practicality, modern taxonomists have amplified the concept of the type specimen far beyond that of the holotypes from which original descriptions are drawn. "Allotypes," "neotypes," "syntypes," and "lectotypes," among many others, can be invoked when a researcher identifying a type encounters some procedural hitch. For instance, when a holotype is lost (which happens occasionally, because of bombing, bad curation of a collection, or other causes), there are rules to govern its replacement by a lectotype, which will now be the "go-to" specimen.

Since Linnaeus didn't designate one, there is and never was a holotype for *Homo sapiens*. Complicating matters is the fact that in the definitive tenth edition of his great work Linnaeus also described the six variants of *Homo sapiens* we mentioned earlier—*Ferus, Americanus, Europaeus, Asiaticus, Afer* (African), and *Monstrosus*. The first and last can be discarded as valid subspecies names, as they do not describe real specimens (*Ferus* was used to designate feral children, and *Monstrosus* was used to denote mythical people with strange morphologies). Under current taxonomic rules, there is one subspecies missing here, as a result of applying the principle of coordination (article 43 of the ICZN). This states that one subspecies of any subdivided species must bear the species name. What this means is that, if we are to recognize subspecies within *Homo sapiens*, we need to add the subspecies *Homo sapiens sapiens*—logically the subspecies Linnaeus had in mind when he wrote his description—to the other three.

In 1959 William Stearns, a taxonomist writing about Linnaeus's legacy, suggested that Linnaeus himself should be the type specimen for *Homo sapiens sapiens* (the name that under the new rules automatically replaces *H. s. europaeus*). And, given Linnaeus's estimable opinion of himself, it is certainly not out of the question that he had himself in mind when he gave our species its name. Since we don't know for sure who he had in mind as exemplar, though, he would have to be designated as a lectotype to satisfy Stearns's proposal. Of course, this would not help very much, because Linnaeus currently reposes in a churchyard in Uppsala, and it is not really practical for any contemporary taxonomist actually to use him as a standard of comparison. Availability is a key consideration, and technically it became a factor some thirty years after Stearns made his original suggestion, when a group of researchers hoping to honor Edward Drinker Cope (and apparently unaware of Stearns's proposal) suggested that Cope be designated the type specimen for *Homo sapiens*. Cope was a famous paleontologist who had willed his bones to science in the hope that he would be designated the type specimen (again, technically the lectotype) of *Homo sapiens*. And it certainly seems that Cope himself had wanted that: a possibly apocryphal story goes that a visitor to Cope's laboratory shortly after the latter's death in 1897 found his long-time technician weeping in front of a boiling preparation vat as Cope's head periodically bobbed to the surface.

For anyone who cared, having two pretenders to the status of type specimen for *Homo sapiens* created a legally awkward situation that demanded resolution. Fortunately, the ICZN was up to the job. One rule states that a neotype can be assigned to a specimen if the lectotype is lost; and this might have given Cope's bones a fighting chance as a neotype. But while Linnaeus is not exactly available, he is not exactly lost, for we know where he is. And his claim is reinforced by a couple of other provisions of the ICZN. One of them is that article 74.1 of the code happens to require that any lectotype must be among the specimens examined by the person who named the species. Linnaeus was long dead when Cope was born in 1840, so Cope could not have been "examined" by the namer. Also, the key article (74.1.1) clearly states the principle of priority under which validly proposed earlier names trump names put forward later. So unless Linnaeus's name for our species is somehow deemed invalid—which, even in an uncertain world, is not going to happen—Cope's claim doesn't stand a chance. Meanwhile, *Homo sapiens* still lacks a usable type specimen.

This digression nicely illustrates the fact that, while systematists will always legitimately disagree, nomenclature is underpinned by an objective set of rules that we must apply in retrospect to Linnaeus's four nonimaginary variants of *Homo sapiens*. The Swedish savant used highly typological reasoning to come up with what we would have to call subspecies, although he probably thought of them as races. And typologically, Linnaeus clearly felt justified in designating his four "real" subspecies based on geography and the skin color of the people involved. He also added some behavioral descriptions that he felt were diagnostic. These were pretty much in line with common European suppositions of the eighteenth century, and paramount were the ways in which the various groups controlled their behaviors. The "rufus" (red) *Homo sapiens americanus* from the New World used custom to govern its behavior; the "albus" (white) *europaeus* was governed by laws; the "luridus" (sallow) *asiaticus* from Asia was opinionated; and the "niger" (black) *afer* from Africa was impulsive. This blatantly racist and typological view of humans was hardly unusual for its time, and it remains significant as one of the first attempts to systematize the differences between human geographic groups.

In technical terms, Linnaeus's trinomina stood until 1825, when Jean-Baptiste Bory de St. Vincent decided to elevate the subspecies names to species level and to add a raft of other regional populations to the genus *Homo* as separate species. But—to cut short a very long story that you can read about in our *Race? Debunking a Scientific Myth*—later experts have synonymized all of these with *Homo sapiens*, so that today no living *Homo sapiens* subspecies are recognized. Even *Homo sapiens sapiens* is entirely superfluous, since we have nothing to distinguish it from. Today, then, while we must revere Linnaeus for his achievements as a taxonomist, we must also admit that his splitting of the species *Homo sapiens* into geographic subspecies was the start of a hugely problematic—and hard to reverse or stamp out—trend toward the formal classification of human individuals and populations. Our colleague Jon Marks has suggested that this desire for racial classification has had an even greater impact on our modern life than the binominal system itself.

5

• • • •

DNA Fingerprinting and Barcoding

On January 15, 2009, Captain Chesley (Sully) Sullenberger crash-landed US Airways Flight 1549 on the almost icy surface of the Hudson River in New York City. Captain Sullenberger's uncommon cool nerves and quick thinking have become legendary; but what caused the crash? From examination of the engines of the jet, it was obvious that a bird strike had occurred; but this unusual event had happened in midwinter, when migratory birds are rare in the skies around New York City. Obviously, if a bird (or any animal for that matter) gets sucked into a jet engine there won't be much left over to examine. Maybe some blood, or some skin, or perhaps the shaft of a feather smeared onto an engine part; but not much else. This is where science comes in; and in a recent paper, researchers at the Feather Identification Laboratory (FIL), one of the more obscure labs at the Smithsonian Institution, reported that they had been able to determine just which kind of bird had caused the calamitous loss of engine power that caused Flight 1549's crisis.

The researchers at the FIL used a then-new technique called "DNA barcoding" to identify the species to which the bloodstains and feather fragments found in the engines belonged. Developed in 2003 by the Canadian scientist Paul Hebert, DNA barcoding uses a short reference DNA sequence known as the cytochrome oxidase I gene (COI) to identify tiny DNA samples to the species level. Hebert had recognized that DNA sequences from this gene harbored enough variation among species to be helpful in identifying them, but that making them useful for this purpose required a database of COI sequences. This led him to establish the Barcode of Life Database (BoLD) as a repository of sequences for all animal species on the planet. BoLD has proven to be an essential source in allowing query sequences from unidentified specimens to be cross-compared with known reference sequences. When the bird remains in Sullenberger's failed engines were examined using this approach, the database made it possible to identify, with 100 percent accuracy, that the unlucky birds involved in the strike were from the species commonly known as the Canada goose (*Branta canadensis*).

DNA barcoding has huge forensic importance when applied to biological materials that cannot be identified from the morphology of a sample. The technique has been used to examine everything from the ingredients in sushi and caviar, to the components of herbal extracts, to illegally imported ivory products and bushmeat. But it is also useful beyond forensics, since it allows the identification of species in many other biological, ecological, and evolutionary contexts. For instance, in collaboration with entomologists Dan Janzen and Winnie Hallwachs, Hebert determined that what was once thought to be a single species of butterfly was in reality ten different ones. Because it can be highly automated, the barcoding approach also promises to speed up the identification of species in the long-term and wide-ranging ecological surveys that are becoming increasingly popular. Identifying all the species from a single bag of insects or nematodes or spiders under a microscope might take a team of morphologists weeks to complete; but DNA barcoding can accomplish the same thing in a fraction of that time.

If all of this sounds familiar to those who like to watch TV crime scene shows, it should. For the very same principles that are involved in the "DNA fingerprinting" beloved of TV writers are also used in DNA barcoding. Both techniques start with DNA sequences, and both query a database for matches. But while BoLD is used by DNA barcoders to make matches to species, DNA fingerprinting attempts to identify a single individual. Several databases are available for accomplishing this. The FBI in the United States maintains a database called CODIS (Combined DNA Index System; figure 5.1); the U.S. armed forces now routinely collect DNA from service members; the United Kingdom maintains the National DNA Database; and

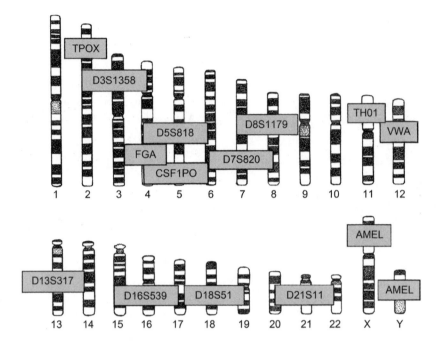

Figure 5.1 Diagram showing the original thirteen CODIS microsatellite regions on the chromosomes of the human genome. The labels give the codes for the most-often used loci in most DNA fingerprinting approaches. Redrawn from the National Institute of Standards and Technology website (http://strbase.nist.gov/fbicore.htm).

many other countries in Europe also have DNA databases. Recently, direct-to-consumer outfits have emerged in the private sector that also do a kind of DNA fingerprinting, and we will have a lot to say about these later. At this point, though, we should point out that DNA fingerprinting does not always need a database to work. If an individual's fingerprint is not in a database, investigators can take DNA from both a crime scene and a suspect and see whether the samples match.

DNA fingerprinting depends on the fact that, while compared with most other species humans are an incredibly closely related group, we nonetheless show a lot of genomic variability among ourselves. On average, two unrelated individuals will have three million base-pair differences—although this means that any two randomly chosen individuals are still about 99.9 percent similar. We must also keep in mind that this is an average difference, and that no single gene sequence or polymorphism can identify any individual human. Establishing the probability that a DNA sample comes from a specific person thus becomes a combinatorial process, necessarily aggregating information from several genes or gene regions.

If a crime scene contains an object that bears DNA, that DNA can be isolated. Once this is done, the DNA can be analyzed using several approaches. Most accurate is comprehensive sequencing of the DNA at the genome level. But because of the expense involved, this approach is not used in most forensic cases. Instead, rapid shortcut approaches are most commonly employed. These take a small subset of nuclear DNA segments (in the past thirteen segments, and now twenty: see figure 5.2) called "loci." These markers are all polymorphic among humans and are known as "length polymorphisms," "microsatellites," or "STRs" (short tandem repeats). They

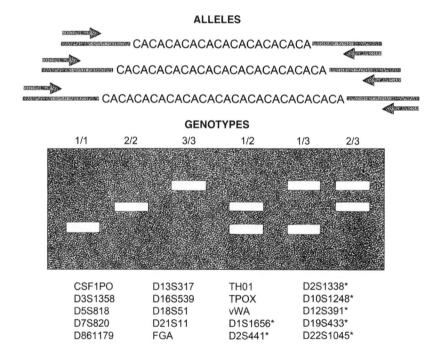

Figure 5.2 Three microsatellite alleles and how they are detected using PCR and gel electrophoresis (rectangle). The list at the bottom is a list of the twenty currently recognized microsatellite loci used by CODIS. Asterisks indicate the seven new microsatellites added on January 1, 2017. Redrawn from the National Institute of Standards and Technology website (http://strbase.nist.gov/fbicore.htm).

are caused by variation in the numbers of bases at specific gene regions in the human genome. Figure 5.2 illustrates the possible states that a given microsatellite might show.

Microsatellites are caused by simple repeats in regions of the genome. In the figure, the repeat concerned is a simple two-base (dinucleotide) repetition of "CA." The repeated units can be any length, but are most commonly di-, tri- and tetra-nucleotides. The top repeat is eleven CAs long, the middle is thirteen, and the bottom is fifteen. The arrows represent conserved regions flanking the repeated region.

Shortly after its invention in the mid-1980s, the polymerase chain reaction (PCR) became the linchpin of modern molecular biology. Taking advantage of the way in which DNA replicates, PCR is ingeniously simple and has very few "moving parts." It simply requires two conserved regions that flank the DNA region of interest. The conserved regions on the right and left of the variable region are used as "primers" for the PCR, which then makes multiple copies of the region of interest between them. These copies can then be visualized using standard molecular techniques. When PCR is accomplished (the arrows in the figure indicate the points that bound the reactions), the three microsatellites will yield differently sized products.

When a suspect is identified in a crime investigation, a sample of tissue (usually a mouth swab) is taken. DNA is then isolated from the sample, which is analyzed for the twenty microsatellites. The polymorphic states for the suspect and the evidence found at a crime scene are then compared; if they match, the probability that the DNA at the crime scene comes from someone other than the suspect can be computed. The various alleles used in forensic DNA profiling have been selected because there is abundant information about them. The definitive frequencies for the thirteen microsatellites used before 2017 and the seven added on January 1, 2017, are published in scientific journals and are used in the probability calculations.

Each of these microsatellites is what is called polymorphic; in other words, it has multiple forms or alleles. The alleles occur in populations at different frequencies that can be found in the literature as frequency tables. Once the profile of the microsatellite alleles of the suspect that match with the evidence has been determined, the frequencies at which those alleles exist in populations are then multiplied by one another to give the final probability that the DNA from the evidence comes from someone else,

TABLE 5.1
An Individual's DNA Matches to Thirteen Microsatellite
Alleles and the Population Frequencies for Those Alleles

Microsatellite	Frequency
CSF1PO	0.13
D3S1358	0.22
D5S818	0.05
D7S820	0.15
D8S1179	0.25
D13S317	0.33
D16S539	0.26
D18S51	0.14
D21S11	0.21
TH01	0.27
TPOX	0.33
vWA	0.06

and not the suspect. Table 5.1 gives an example of an individual's DNA matches to thirteen alleles and the population frequencies for those alleles.

To determine the probability that one could find another person with the same alleles, you just multiply the frequencies together

$$(0.13)(0.22)(0.05)(0.15)(0.25)(0.33)(0.26)(0.14)(0.21)(0.27)(0.33)(0.06) = 0.000000000723154$$

or

$$7.23154 \times 10^{-10}$$

or

1 in 7,231,540,000

or

about 1 in 7 billion

In this instance, which uses the old CODIS system of thirteen microsatellites, we see the frequencies of alleles that match in an imaginary case. The overall probability that the crime scene DNA comes from someone other than the suspect is about one in seven billion. This is a very low probability indeed, given that there are a mere seven billion people on the planet. Tests like these are neither easy nor inexpensive to do, nor should they be done by amateurs. But a multitude of commercially available microsatellite kits—with names like Profiler, Identifiler, Globalfiler, and PowerPlex—are registered and approved for DNA-profiling labs to use. And not just law enforcement labs put them to use. This is the same procedure that Maury Povich uses on his TV show to determine paternity, and it is how the DNA laboratory of the Office of the New York City Medical Examiner went about identifying the remains of the 9/11 World Trade Center victims.

In addition to STR profiling, gender profiling and mitochondrial DNA (mtDNA) profiling can be used to determine both the gender and the maternal lineage of an individual. Gender profiling takes advantage of the fact that males have a Y chromosome and females do not. To tell if a sample comes from a male, one need only probe the sample with markers for the Y chromosome. If the sample contains DNA from a female, no product will be engendered. The mtDNA profile is important because, as we will shortly see, the mtDNA lineages of most major ethnic groups on the planet have been determined, providing a database against which anyone's DNA can be compared. This database is what a lot of direct-to-consumer ancestry companies consult to tell customers where they come from—which is another issue we will address in later chapters.

Despite the similarities between DNA fingerprinting and barcoding, the two differ fundamentally in that barcoding uses diagnostics, whereas profiling uses allelic frequencies. They are also implemented rather differently. DNA barcoding is implemented in two major ways, via phylogenetic trees and diagnostic characters. Both approaches require a database (BoLD), but they use it in different ways. In the tree-based approach, the closest relatives to the specimen of interest (the "query specimen") are extracted from the database, and the query specimen is added to them. A phylogenetic tree is constructed using any of the methods we described in chapter 3, and the position of the query sequence in the subsequent tree tells the researcher its identity. Most of the time the query sequence will show up as a sister taxon to one of the database sequences, allowing the researcher to assign it to the same species. In some cases, however, the phylogenetic analysis might place the query in an unexpected place relative to the database sequences. In this case, the researcher has two options. The first is to conclude that maybe the database is incomplete and to increase sampling accordingly. Alternatively, the researcher can hypothesize that the query specimen represents a new species.

The second approach uses a method called "population aggregation analysis." In this approach the same sequences used in the tree analysis are used. The difference is that the sequences from the database are aggregated before the query sequence is examined. In figure 5.3 we see how this is accomplished for a short example sequence of three species with five individuals each. Each species has a single base that is diagnostic for the species (in caps and bold type). The query specimen is identified as a member of species 2, because it has an "a" in the sixth position: the diagnostic state for species 2.

Species 1
Individual 1	agTctactgacgt
Individual 2	agTctactgacgt
Individual 3	agTctactgacgt
Individual 4	agTctactgacgt
Individual 5	agTctactgacgt

Species 2
Individual 1	agactTctgacgt
Individual 2	agactTctgacgt
Individual 3	agactTctgacgt
Individual 4	agactTctgacgt
Individual 5	agactTctgacgt

Species 3
Individual 1	agactactgacTt
Individual 2	agactactgacTt
Individual 3	agactactgacTt
Individual 4	agactactgacTt
Individual 5	agactactgacTt

Query   agacttctgacgt

Figure 5.3 Using population aggregation to identify a query sequence. The bold capital letters indicate the base in the short sequence that is the diagnostic. The bold lowercase letter in the query indicates the diagnostic position and change that allows the designation of the query as a member of species 2.

Note that the diagnostics for the three species are all fixed, and differ from the others, so the query sequence simply needs to be scanned for the presence or absence of these diagnostics. For instance, if the query sequence has a "t" in position 3, it can be diagnosed as species 1 provided the rest of the sequence does not have the diagnostics of species 2 and species 3. In this example, the query sequence is identified as species 2 due to the presence of the species 2 diagnostic only. But what happens when the query sequence is agTctTctgacTt, and thus shows diagnostics from two or more species (in this case the query sequence has three diagnostics)? The query sequence then becomes one of those wonders of science (an exception to the rule) and will disrupt the diagnostic system of the three species. One way out of the disruption is to sequence more of the specimens in the three species and establish additional diagnostics. If you are lucky, the new sequences will harbor unique diagnostics, and you can proceed using them. Alternatively, you could go back to where you found the query specimen and collect more individuals in that area in the hope that you might be able to diagnose a new species that includes the query specimen. If this sounds familiar, it is. The same procedure is used in the classical taxonomy that we described in chapter 4, but the procedure we are discussing here uses DNA sequences.

Which is the better way to do DNA barcoding? This is an active question in systematics and taxonomy. Our preference is to use the character-based approach to find diagnostics, because it is in line with the way taxonomy has been done for two hundred years. We should look on DNA as a valid source of characters to describe species, just like morphology. Including DNA along with morphology in this way is known as "integrated taxonomy." In contrast, using the phylogenetic tree approach in barcoding is problematic, not only because taxonomists have in general refrained from using trees to do taxonomy at the species level, but because it has an objectivity problem since deciding what a sufficiently divergent monophyletic group is can be highly subjective. Remember that a monophyletic group is a set of taxa that are each other's closest relatives, to the exclusion of all other taxa. With a phylogenetic tree, it is an arbitrary thing to pick a node above which one can call all individuals members of the same species. DNA barcoders use what they call a "barcoding gap," a point at which they perceive an excessive genetic distance separating adjacent individuals or taxa; but again, such distances are arbitrary and can change with the kind of organism being studied. Nonetheless, both approaches are routinely used to identify species and query sequences.

Both DNA profiling and barcoding are relevant to our discussion of race in humans for several reasons. It is possible, for example, to take all the sequences in the database from the various populations on the globe that are thought to be races and use them in a DNA barcoding or profiling exercise. But if we tried this approach, we would rapidly encounter problems. One important issue involves the assumptions underlying both DNA barcoding and DNA profiling. Both approaches are made operational by assuming group inclusion, so deriving any inferences about group inclusion using either of them would involve circular reasoning. So making any application of these approaches to race scientific would require adopting a hypothesis-testing approach; and as we discuss next, there are insurmountable obstacles to creating the testable hypotheses needed to scientifically validate any concept of human races.

6

• • • •

Early Biological Notions of
Human Divergence

Linnaeus was the first systematist to formalize what many Europeans saw as major differences between people in Europe and people from other geographical regions. He did this primarily in the service of situating humans within the context of the amazing variety of life on the planet; but as we've seen, there is equally no doubt that his classification and the remarks accompanying it also reflected the stereotypical and implicitly racist attitudes of his time. Most fatefully of all, though, in recognizing his four varieties of *Homo sapiens*, Linnaeus opened the door to a tradition of classifying our species into formal (and therefore separate) groups that became ever more blatantly agenda-serving as time passed.

The most influential of Linnaeus's immediate successors in this area was Johann Friedrich Blumenbach, a German scholar who, oddly enough, spent an unusually sheltered career at the University of Göttingen. First published in 1776, Blumenbach's thesis *On the Natural Varieties of Humankind* followed Linnaeus in recognizing four human "principal varieties," although by the time of the third edition of 1795 the number was up to five: the Caucasians (Europeans), Mongolians (Asians), Ethiopians (Africans), Malays (Indo-Pacific inhabitants), and Americans. Significantly, Blumenbach recognized "insensible transitions" among these varieties, and explicitly accepted those transitions as evidence that all living humans belonged to a single species. But his notion that *Homo sapiens* was divided into five major geographic/racial groups (variously defined) later hardened into dogma.

Alas, the early nineteenth century, the golden age of slavery, brought with it more overtly political agendas. Explicit theories of racial inferiority began to emerge, and debate began to center around whether the various races had a single origin (monogenism) or had been separately created (polygenism). And once Darwin had gone public with his evolutionary ideas in 1859 with the publication of the *Origin*, the increasingly vocal advocates of black inferiority began to proclaim that the races had different ancestors.

This appalled the monogenist and abolitionist Darwin, whose views of humankind had been deeply affected by the brutality and cruelty of slavery he witnessed in Brazil during the voyage of the *Beagle*. Indeed, four decades after this experience, his masterful *Descent of Man* was in large part a plea for the monogenist cause. Yet Darwin was also able to write in the *Descent* that "the Negro and the European, are so distinct that, if specimens had been brought to a naturalist *without any further information* [emphasis added], they would undoubtedly have been considered as good and true species" (Darwin 1871, 217).

Of course, human beings are more exquisitely attuned to variations within their own species than within any other, but Darwin's statement is nonetheless symptomatic of a real dilemma that evolutionary biologists and naturalists face when considering race and the differentiation of humans. Darwin's caveat makes it clear that what was engaging his attention here was not the question of species status but the sheer amount of variety that can accumulate within a single species. And however superficial the differences among humans may in fact be, it is nonetheless undeniable that they are striking. Our favorite modern-day illustration of this is the famous photograph showing Christina Aguilera standing between basketball legends Shaquille O'Neal and Yao Ming (google "Shaq Yao Christina" if you're curious). It shows just how eye-catching the differences among humans can be, and it tells you what Darwin was thinking of when he wrote the words we just quoted.

The key to understanding Darwin's scientific point, though, lies in the phrase "without any further information." Because, as we've seen, species are not just what they look like. We need to consider a lot of other evidence in weighing whether—or not—the visible or even the molecular (DNA) differentiation among populations of humans (or populations of any other organisms) also reflects the evolutionary separation of lineages. Of course, if you stand in the middle of New York City's Times Square, it is hard to resist the temptation to guess the geographic origins of the people passing you by, based on their appearance. But how accurate would you be? And, just as importantly, would your notions be the same as those of others standing near you doing the same thing? Chances are that in many cases you would be way off, maybe even getting the continent wrong. And we also suggest that, if you compared notes with your neighbor, the impressions you formed would turn out to be very different from his or hers—at least partly because you would more than likely be using different criteria. External appearance, then, is a tricky thing; but might we have more success with skeletal or fossil remains, or with molecular data?

We have only a scattering of fossils of early modern humans, and using them to interpret the early movement of *Homo sapiens* across the globe is a hazardous business. Take the example of "Kennewick Man," the nine thousand-year-old skeleton found on the banks of the Columbia River in Washington State in 1996. In terms of preservation, this skeleton is about as good as it gets in the fossil record; but the story of its interpretation is nonetheless a tortuous one. It was originally studied by forensic experts, who noted the presence of "Caucasoid traits [and a] lack of definitive Native-American characteristics" (Chatters 1997, 9). This was odd, to say the least, especially given that embedded in the pelvis was a stone point of the kind typically made by the Paleoindian populations of the period. But a reconstruction of the individual in life apparently showed an eerie resemblance to the English actor Patrick Stewart, familiar to us in his TV role as captain of the starship *Enterprise*!

The mystery deepened a few years later, when other experts, using cranial metrics, concluded that the closest comparison of the Kennewick skull was not with American Indians or Caucasoids, but with the Ainu, descendants of early settlers of the Japanese archipelago. Could the Kennewick skeleton have been the remains of an ancient Pacific seafarer who had come all the way to the western United States only to be speared in the pelvis by an angry local? In the end, the matter was settled by DNA studies that were finally made possible by advancing technology. These showed that the true story of Kennewick Man was the one that had been most likely from the start. Genomically, the Kennewick fossil most closely resembles living Native Americans. A reluctance on the part of local Native Americans to give DNA samples has made comparisons difficult, but the latest assessment is that Kennewick Man's genetic profile is particularly close to those of modern members of the nearby Confederated Tribes of the Colville Reservation. Little matter; the DNA evidence, which was much more straightforward to interpret, had proven much better than morphology in determining the fossil's affinities.

No wonder, then, that the signal emerging from the usually much more broken-up and less completely preserved fossil record of early *Homo sapiens* is far from clear. Our species is a very young one, so for all its vast distribution, it is also very closely knit; and other fossil bones have accordingly proven just as hard to interpret as Kennewick Man or, for all their superficial variety, those folks in Times Square. But it has taken a long time to reach this realization, and some forensic investigators still feel strongly that they can visually determine the "race" of skeletal remains. This forensic enterprise is a hazardous one; and although someone knowledgeable can usually make a reasonable guess at whether the possessor of a skull dug up in the woods was "black" or "white" or "Asian," quite often they will be stumped. Or if they think they know, they could easily be just plain wrong. So most forensic scientists are very careful about assigning possible racial origin to skeletal remains based entirely on the morphology of the bones.

We will look later at how DNA has been used to trace the differentiation of our species and its movements across the face of the planet. Meanwhile, we should at least glance at various questions of morphology as they relate to human variety, since the significant problems surrounding a morphological approach to determining race haven't prevented a whole lot of researchers from attempting to categorize skeletal remains by race or, conversely, to examine the morphology of people in different so-called races. Perhaps the most famous study of human cranial variation was undertaken by the nineteenth-century Philadelphia naturalist Samuel George Morton, whose conclusions were famously analyzed a few decades ago by Stephen Jay Gould and have more recently been reviewed by John Michael and by Jason Lewis and colleagues. As a product of his time, Morton held highly racist views on human variety, and in his day he was most famous for his "American Golgotha," a collection of more than one thousand human skulls that formed the basis for his research. The skulls came from all over the world and represented a wide range of what at the time were considered racial groups.

One of Morton's uses of this collection was to measure the cranial capacities of individuals classified in different racial groups in the attempt to correlate race with intelligence. Gould's reanalysis of Morton's data centered on the possibility that Morton had fudged his data to reach the conclusion that white people had bigger brains than other people on the planet. Lewis and colleagues closely reanalyzed Morton's original specimens, and, like Michael, found little if any reason to identify deliberate hanky-panky in Morton's analysis. The ordering of cranial capacity of the "racial" groups Morton recognized changed a bit with reanalysis, but on average, skulls from "white" people nonetheless still tended to be a bit larger than other people's skulls. Nonetheless, Lewis and colleagues found many problems with Morton's approach and procedures, a primary one being that the various "racial" clusters accepted at the time showed so much overlap in cranial capacity as to render the analysis meaningless. In other words, the variation within Morton's predetermined groups was just so large that significant overlap between the groups was inevitable. Second, because males are larger bodied, they have bigger brains than females; and the relative numbers of males and females in each group also affected the outcome of the analysis. Finally, there is no neurobiological reason to think that a larger brain necessarily means higher intelligence. After all, both Neanderthal and early *Homo sapiens* skulls have cranial capacities on average 13 percent larger than those of living *Homo sapiens*. And we would guess that what is true for cranial capacity is also true for the rest of the skeletal characters that have been examined for racial differences.

More recent researchers have focused on external, physiological, or behavioral differences among people on our planet. But here we need to realize that humans are biological creatures, the outcome of a long and intricate process of negotiation between the genome and the environment that surrounds it. Even if you identify a difference that looks interesting to you, and you want to ask why people here are different in this trait compared with people over there, you will always run into the problem of whether the trait is genetic, acquired, or somewhere in between. If it *is* genetic, you are also faced with the problem of how complex the genetics of the trait are. Some traits might be inherited in a simple fashion, but others might involve hundreds of gene interactions and complicated gene regulation. Features such as intelligence, athletic ability, dietary preferences, skin color, and body shape are all complex traits that have been attributed to racial groups. But in the event, none of these works well at all in diagnosing discrete human groups on the planet.

Part of the problem is that almost none of the classic physical factors by which different regional human groups have commonly been recognized appears to have much adaptive significance. Take the epicanthic folds that are commonly seen on the eyelids of people of eastern Asian origin. We have no idea why those folds are there. From what we know about the demographics of human expansion into Asia, it is obvious that the trait initially attained a high frequency in the founding population—but it did so almost certainly because of genetic drift in that very small population, rather than because it offered any physiological advantage. The eyefold is, indeed, emblematic of the many neutral traits that stick around in human populations because they simply don't get in the way of survival.

The most striking exception to the notion that most interpopulation heritable variations among humans are nonadaptive is skin color. Here a pretty convincing case for adaptation can be made, if not under present circumstances, then in the past. Our skin is the largest organ of our bodies, and it serves as the first line of defense from outside invasion by other organisms, chemicals, and molecules. Low-latitude environments such as the ones in which our African ancestors resided are tough on the skin because of the destructive ultraviolet component in solar radiation. Hair is one way to protect the skin from ultraviolet radiation (UVR), and in the absence of hair the same function can be performed by the dark pigment melanin, which very efficiently absorbs around 99.9 percent of all the UVR that reaches it.

Just when our ancestors lost their ancestral hairy coat is not entirely clear. But it seems reasonable to suppose that the process began at around two million years ago, when those ancestors first moved out of the shelter of the forest and woodlands and committed themselves to life out on the sun-soaked African savannas. The shadeless new environment made the control of body heat a critical thing, for even momentary overheating of the brain can have dreadful physiological effects. Early humans came up with the expedient of cooling the body—and hence the brain—by the evaporation of sweat produced by cutaneous glands. This vital process would have been impeded by a substantial coat of body hair; but losing that hair meant exposing the delicate skin to the sun's rays, and that would have placed the protection that melanin affords at a premium. Which is why the early members of our tropical genus *Homo* almost certainly had dark skins, though whether they were "black" or "bronze" is entirely a matter of speculation.

Today, of course, vast variation in skin pigmentation exists both within and between human populations from across the globe. So how did this variation come about? Well, it turns out that the skin's relationship to UVR is complex, because UVR both inhibits and stimulates the production of chemical products that maintain physiological balance in the body. Folate, a vitamin essential to DNA replication in cells, is destroyed by longer-wavelength UVR, or UVA. On the other hand, the shorter-wavelength radiation (UVB) promotes the synthesis in the skin of vitamin D (needed for proper calcium metabolism). These two kinds of UVR come in a package, and in her book *Skin,* the anthropologist Nina Jablonski has recently emphasized the necessity of balancing the inhibitive and stimulating effects of UVR upon the skin. Maintaining this balance has led to two distinct gradients of environmental influence. One cline runs from the poles to the equator, while the second goes in the opposite direction; but both favor darker skin in the tropics and lighter skin at higher latitudes, for as dark-skinned early modern humans migrated out of Africa they encountered environments in which high melanin content in the skin was not strongly selected.

In outlining scenarios of this kind, it is difficult not to fall into the adaptationist trap to which our reductionist human minds seem innately attracted. But while it is obvious that heavy pigmentation is adaptive in the tropics, the dynamics of higher latitudes are not as clear. Perhaps there is simply a physiological or other cost to maintaining high melanin levels in the skin in higher latitudes, in which case selection would tend to move toward elimination of deep pigmentation. Or maybe vitamin D synthesis looms larger in the equation. Or perhaps what we are seeing is simply the result of relaxed selection for dark skin pigmentation at higher latitudes. Nonetheless, the tendency is there.

The molecular basis of skin color variation in humans (and in Neanderthals, for that matter) provides us with an excellent example of opportunistic nature at work. The gene at the heart of current research on skin color is known as SLC24A5. This gene produces an enzyme that shows great variation in its sequence among populations around the world, and teasing apart what this variation means with respect to skin color and adaptation is not simple. Almost all Africans have alanine in a key spot in the protein the gene produces, while as many Europeans have threonine in this key spot. The alanine and threonine make proteins that behave differently, accounting for much of the average difference in skin color between people from the two continents. If Europeans and Africans were the only *Homo sapiens* on the planet, the story of skin color would therefore appear very simple. However, it turns out that that 93 percent of (dark) Africans and (much lighter-skinned) East Asians share the alanine in that key spot. Europeans and East Asians thus show apparently similar light skin via different genetic pathways, so the dark-skinned tropical ancestor of our species must have given rise to the lighter-skinned, higher-latitude Asian and European skin patterns via different mechanisms. Don't forget, either, that this dark-skinned African tropical ancestor also gave rise to the current dark-skinned populations elsewhere on the planet.

In the same set of studies that turned up this information, researchers examined the genetic architecture of skin color in a population of people of mixed African and European ancestry, among whom both the alanine and threonine types of melanin exist in high frequencies. The study genotyped individuals and measured the quantity of melanin in their skins. People homozygous for the alanine type had the most melanin in their skin; those individuals with two copies of the threonine type had less melanin in their skin. Again, it is difficult not to fall into the "simple story" trap, at least until we look at the heterozygotes and the distribution of homozygotes more closely. It turns out that only 38 percent of the variance in skin color can be accounted for by the melanin gene in question. The remaining 62 percent of the variance must be attributed to other gene complexes or to genetic mechanisms other than simple Mendelian genetics. The upshot is that there are not only many ways in which to have skin with low quantities of melanin, and many ways in which to have skin with high amounts of melanin, but even more ways in which in-between amounts of melanin in skin are produced.

It seems increasingly clear that all attempts made before the advent of genomics to interpret the significance of "racial" traits are hugely questionable. Not that this inhibited lavish discussion of "racial differences," often through the expedient of making phylogenetic trees. Yet while we have seen that hierarchy is an essential tool in any attempt to understand our larger historical context, it is much less evident that it has much relevance to understanding historical divergence within our species. We will return to this issue later; but at this point it is relevant to look at the patterns of hierarchy inferred by researchers in the days before the genomics revolution. Those patterns were derived using the "evolutionary taxonomy" approach, in which trees were drawn based on the authoritative pronouncement of those who drew them; and while such trees were in some way based on data that the scientist felt were important, they were not constructed using an algorithm or any optimality criterion.

The systematist David Morrison has created a scholarly website called the Genealogical World of Phylogenetic Networks (http://phylonetworks. blogspot.com), where he and colleagues chronicle the development of tree thinking and tree construction. According to the authors of the website, trees have been used since the beginning of the nineteenth century to depict the relationships of human populations to each other. As far back as 1800, Felix Gallet used the characteristics of language to construct a very tree-like diagram of human group relationships (figure 6.1), the first of many language-based attempts to reconstruct the history of human populations. Still, it is important to remember that languages are cultural constructs that can be acquired and do not necessarily reflect the biological relationships of the groups involved.

One of the first attempts to represent human "racial group" biological relationships was made by Sir Arthur Keith in 1915 (figure 6.2) and shows branching patterns within the genus *Homo*. It depicts the divergence of the Neanderthal lineage between a very broadly estimated nine hundred thousand and four hundred thousand years ago and shows the four major lineages of modern humans considered critical at the time—African, Australian, Mongoloid, and Caucasian—branching at around three hundred thousand years ago.

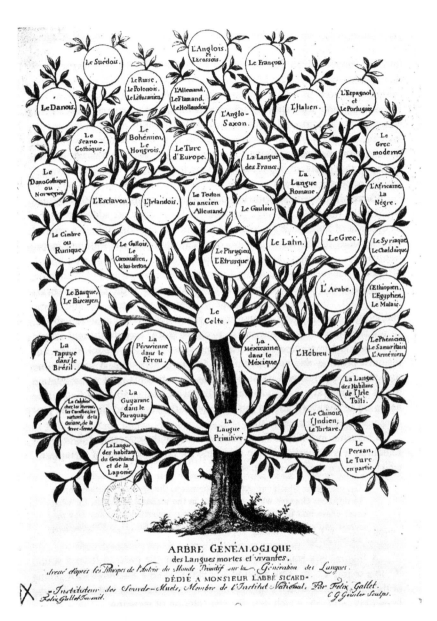

Figure 6.1 Gallet's (1800) tree of human languages.

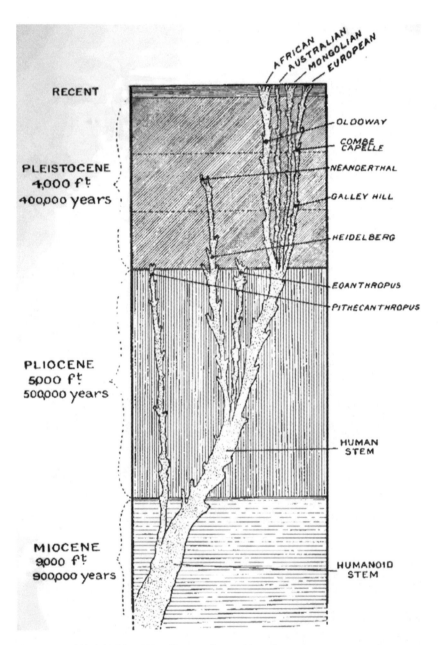

Figure 6.2 Keith's (1915) branching diagram of genus *Homo* lineages.

Significantly, Keith made no attempt to show any branching pattern within *Homo sapiens*, deriving all groups from the same time point.

Perhaps the most visually spectacular representation of human relatedness came from John B. Sparks, who was famous for drawing what he liked to call "histomaps." Using published information, he produced beautiful diagrams that represented historical developments. While he did this for all of life, in one histomap he did the same for religions and world history. This 1932 histomap (see plate 2) is spectacular in its ambition to show, in a single diagram, how all life on the planet diversified; and the bottom one-third of the diagram shows the patterns of supposed emergence among human groups. It is incredibly Eurocentric, but admirable in its attempt to boil everything down to common ancestors and subsequent divergence from them.

Paleontologists of the 1940s had different ideas about the divergence of modern human lineages. Earnest Hooton and Franz Weidenreich both published diagrams (figure 6.3) of human branching based on paleontological data. Not surprisingly, since they focused on the same kind of data, their depictions broadly coincide. Their diagrams look very different, as Hooton chose a metaphorical circulatory system to depict relationships,

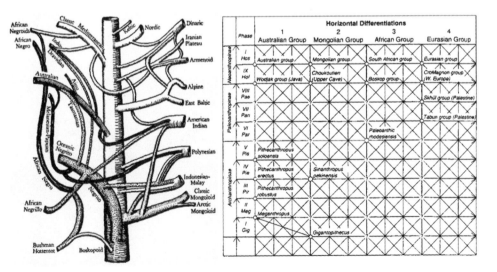

Figure 6.3 Hooton's (1946; *left*) and Weidenreich's (1946; *right*) branching drawings showing the lattice-like interpretation of the relationships of human geographic groups.

and Weidenreich imposed a rigid grid. But both diagrams tell the same story of interweaving lineages due to the absence of reproductive isolation.

Modern attempts to represent the relationships of *Homo sapiens* lineages are rather more diverse in concept. As we will see, some researchers today use bifurcating and fully resolved trees; some use reticulating diagrams; and others use a combination of both. Our colleague Olivier Rieppel has neatly summarized this schizophrenic situation: "The history of biological systematics documents a continuing tension between classifications in terms of nested hierarchies congruent with branching diagrams (the 'Tree of Life') versus reticulated relations" (Rieppel 2010, 475).

7

• • • •

Mitochondrial Eve and
Y-Chromosome Adam

Karl Landsteiner was a dedicated scientist who literally died at his lab bench in 1943. A German biologist who can arguably be called the father of molecular genetics, he studied blood groups for the greater part of his career. Scientists before Landsteiner had recognized that when a human was given a transfusion of blood from an animal the results were disastrous, because the blood cells clumped together (technically, "blood agglutination" occurred). Similar problems were also seen in some human-to-human transfusions, and Landsteiner became convinced that similar mechanisms were at work in both cases. Eventually, he proposed that there were multiple kinds of human blood determined by genes. Through studying how different people reacted to transfusion, by 1909 Landsteiner had teased apart the genetics of these blood types. He determined that there were three types of human blood—A, B, and O—and that they were specified by different alleles.

The interesting thing about the three alleles is that they create six different genotypes: AA, BB, OO, AB, AO, and BO; but they phenotypically produce just the four blood groups: A, B, AB, and O. The immunology involved is intricate, but most importantly here, it was possible for geneticists to calculate the frequencies of the various alleles in different populations. The first researchers to do this were Ludwig and Hanka Hirschfeld, who published an analysis of the ABO blood group frequencies in "different races" in 1919, with the intention of deciphering the divergence of those "races." The Hirschfelds' choice of circumscribed populations was enlightening (Indian, English, French, Italian, German, Austrian, Bulgarian, Serbian, Greek, Arab, Turk, Russian, Jew, Malagasy, Negro, and Annamese), and already demonstrated the difficulty in choosing appropriate populations to be analyzed. But they did succeed in showing some interesting variations in the frequency of AB alleles among the populations they studied.

The American geneticist Sewall Wright and English biologists J. B. S. Haldane and R. A. Fisher are credited with founding the discipline of population genetics. And they soon recognized the enormous potential that blood groups and their frequencies offered in studying the genetics of human populations and how genes moved around within them (figure 7.1). At the outset of World War II, Fisher and his colleague George Taylor used the results of blood group typing to make one of the first studies of population differences in England and Scotland, and events subsequently moved fast. It was quickly discovered that, while the ABO system is the most prominent (and most studied) blood group system, thirty-five such systems actually exist, all easy to determine with relatively simple immunological tests. Population geneticists used these systems to examine many aspects of the population dynamics of humans, and rapidly experienced all the same problems that we will encounter when we address the frequencies of DNA sequence polymorphisms. Those problems include the difficulty of circumscribing specific populations and of interpreting their gene frequencies in a global context. For instance, the west African Bororo, Peruvian Indians, and the Shompen people of the Nicobar Islands in the Indian Ocean are all are fixed for the O allele, while most other populations on the planet check in at near 50 percent frequency for the O allele. What does this mean with respect to the evolution of these populations? Are they differentiated from the rest of humanity by this genetic quirk?

Blood group frequencies have been used not simply to characterize human populations, but to reconstruct relationships among them. The first attempt to do this was made in 1963 by Luca Cavalli-Sforza and A. W. F. Edwards, who in 1994 expanded their reach to include more populations (figure 7.2). The two trees present some interesting differences; for example, the terminals in the 1963 tree do not all appear in the 1994 tree, while people of Asian ancestry appear as a monophyletic group in the 1963

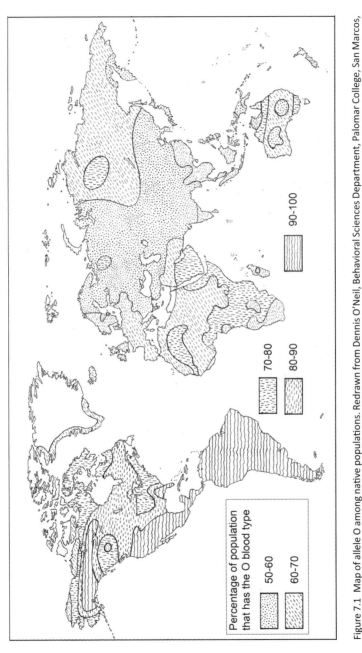

Figure 7.1 Map of allele O among native populations. Redrawn from Dennis O'Neil, Behavioral Sciences Department, Palomar College, San Marcos, California (http://anthro.palomar.edu/vary/vary_3.htm).

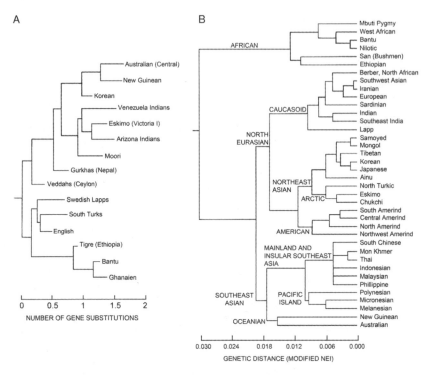

Figure 7.2 A hierarchical representation of fifteen human population groups drawn using five blood group systems and maximum-likelihood approaches. The 1994 tree uses additional molecular data to construct the tree. (A) 1963; (B) 1994. Redrawn from Bodmer (2015).

tree but not in the 1994 tree. The most striking difference of all is how the trees are rooted—something that points to the arbitrary nature of rooting a tree that contains only members of a single species. In the 1963 tree, African populations are most closely related to Europeans, while in the 1994 tree the Africans, while monophyletic, lie at the root. This shift is curious, but it is explained by the fact that the 1994 tree shows African populations at its root because they are the most distant from all other populations in the data set, invoking the assumption that they are more basal in the tree. As we've seen, this kind of assumption is not entirely warranted and objective.

Only blood data were available in 1963, but the 1994 tree included other data, because by that time major new kinds of molecular information were being used for examining both populations and species in an evolutionary context. The first new data source was developed in the mid-1960s and took advantage of the physical properties of the proteins coded for by the genes. There are only 35 blood systems for researchers to work with, but there are thousands of genes in the human genome; and Harry Harris, a British geneticist who spent most of his life studying human variation, tapped into this treasure trove.

In the early 1960s Harris learned of a technique called gel electrophoresis that had been around since the 1930s but was refined in the 1950s. It works as follows: DNA sequences code for proteins; when a mutation occurs in a gene, this will often alter the amino acid sequence of the protein the gene codes for. A change in an amino acid in a protein, even at a single position in the protein, can change its biochemical properties. These can be assayed many ways, but Harris settled on electrophoresis, in which an electrical current applied to a protein moves the protein in the direction of the current. How far the protein moves is a function of the amino acids present. Researchers use gel-like matrices to slow the proteins down and to measure how far they move. By running protein extracts from people on these gels, researchers can discern subtle differences in the proteins involved, and those differences can in turn be tied in to the underlying genes.

Following Harris's work, gel electrophoresis became a mainstay of the study of protein variation in human populations and provided a lot of additional data for Cavalli-Sforza and Edwards's 1994 tree. By the 1970s this "allozyme analysis" had become the workhorse of genetic research in all kinds of populations. Richard Lewontin's classic and aptly titled 1974 book *The Genetic Basis of Evolutionary Change* (though some joked that it should have been called "*The Evolutionary Basis of Genetic Change*") was inspired by his own and his colleague Jack Hubby's allozyme work on fruit flies, but the technique was also widely adopted by those interested in humans, especially the role of rare alleles in human populations. Again, intriguing work came from Cavalli-Sforza and his colleagues. In 1994 Cavalli-Sforza, Menozzi, and Piazza published a fat tome entitled *The History and Geography of Human Genes.* In it the authors delivered, as promised, a large amount of detail about human genes in evolutionary and geographical contexts, and then used these data to make inferences about the relationships among people on the planet. They took data from approximately 40 enzyme systems they had used back in 1988 to generate the tree shown in figure 7.2B and calculated distances among the populations involved. Those distances, not the genes themselves, were then used to produce the tree. But while such procedures are well established in the context of determin relationships among species and higher taxa, we unfortunately encounter a whole host of difficulties when constructing trees in which the terminals represent taxa below the species level—as is the case with human populations.

At about the same time that the allozyme work was building steam, some pioneering researchers were developing techniques to obtain the primary base sequences of genes. The first DNA sequences generated were of viral genomes, and these small sequences of DNA took a relatively long time to generate directly using the techniques available at that time. Accordingly, many scientists adopted a shortcut method—known as restriction fragment length polymorphism, or RFLP—for DNA sequence analysis. The RFLP technique took advantage of the fact that certain enzymes from bacteria could be used to cut DNA at specific four- or six-base stretches. The idea was that if one organism had the sequence GAAG at a certain place in a gene of interest, and another organism had GAGG, the application of an enzyme that cut GAAG (most restriction enzymes cut at symmetrical or "palindromic" sequences) to the gene's DNA would result in the GAAG in the first organism being cut, while the GAGG in the second organism would remain intact. Laboratory methods for detecting these polymorphisms were quickly developed, and it became relatively easy to assay specific genes for the presence or absence of restriction for large numbers of different restriction enzymes, in large numbers of individuals. The process of determining RFLPs for a typical gene was still cumbersome, and a method that focused on an operationally easy DNA target was needed. Fortunately, one existed.

Researchers discovered very early on in DNA studies that the DNA from the mitochondrion (mtDNA) could be much more easily manipulated in the lab than the DNA residing in the nucleus. So, naturally enough, mtDNA was an early operational choice of evolutionary biologists. Nearly every cell in the body has mitochondria, tiny organelles lying outside the nucleus that are often referred to as the "powerhouses of the cell." There are anywhere from one hundred to ten thousand mitochondria in a cell, depending on the tissue concerned. Each mitochondrion has a small, circular, double-stranded genome that, in the human reference sequence, is 16,569 base pairs long. This small genome has thirteen protein-coding genes, twenty-two genes that code for transfer RNAs, and two genes that code for ribosomal RNAs, molecules that assist in the replication process. Because of the large number of copies of this DNA molecule in the tissues of organisms, mtDNA was easy to isolate and became a staple of evolutionary research. Still, the molecule is peculiar. Unlike nuclear DNA, which comes from both parents, it is maternally inherited, passed on from mother to offspring. This is because the mother's egg is a complete cell (with mitochondria), whereas the father's sperm contains only nuclear DNA, so that males are effectively evolutionary dead ends with respect to this molecule. mtDNA is thus clonal, meaning that the genetic material is simply copied and passed along to the next generation, and the only variation that is introduced into this process is caused by mutation. Mutation is hardly uncommon, of course; but mtDNA has turned out to be something of a paradox of variation, as some parts of it evolve very rapidly (i.e., accumulate mutations at a high rate), while other parts of it evolve very slowly (incurring very few changes over time).

In a seminal paper published in 1987 (after forty revisions and more than a year of review by peers) Rebecca Cann, Mark Stoneking, and Allan Wilson analyzed RFLP data for 147 human mitochondrial DNA genomes, representing 133 kinds or "haplotypes" of mtDNA, generating what they called a "genealogical tree" of these 133 haplotypes (figure 7.3). The paper was a scientific blockbuster, and the public was enthralled by the scope of the inferences the authors made. It was this study that spawned the term "mitochondrial Eve," as the work logically implied a coalescence point in the past for all mtDNA haplotypes, representing a common ancestral haplotype for all human populations. Note that we say "common ancestral haplotype," and not "common ancestor," because the terminals in the tree do not represent actual people, but rather haplotypes of mtDNA.

In the thirty years since this pivotal publication, work on mtDNA has exploded. There are literally thousands of human mtDNA studies in the literature. One of the major accomplishments of this research has been to simply organize and categorize the variation in this molecule among humans. Most of the classification of humans into "haplogroups" is done by analyzing two hypervariable regions of a noncoding part of the mtDNA genome

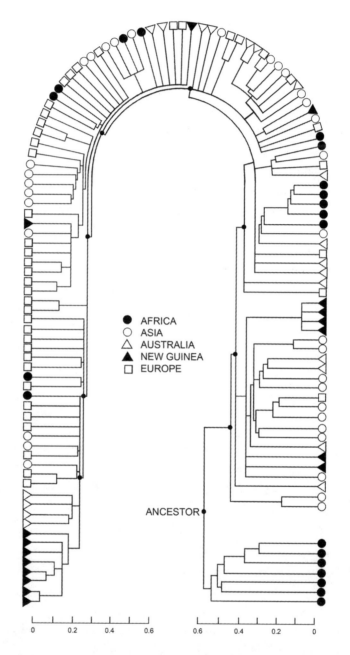

Figure 7.3 Genealogical tree of human mtDNA haplotypes from Cann et al. (1987). The geographic origin of the haplotypes is given in the legend in the middle of the tree. The tree is rooted on the longest branch (also known as the midpoint rooting) of the unrooted distance tree, making African haplotypes the most basal geographic group in the tree. Redrawn from Cann et al. (1987).

called the D-loop. These two regions are called HVR1 (16024–16569) and HVR2 (0001–00576), where the numbers in the parentheses refer to the location of the sequences in the Cambridge Reference Sample (the sequence of an arbitrarily chosen human that is designated the "reference" or "type").

In 2009 Mannis van Oven and Manfred Kayser standardized the haplogroups by generating a phylogenetic tree from the existing data. They used a nomenclatural system for the different haplotypes that is hierarchical but based on letters and numbers. Haplotypes are given a major monophyletic group (clade) designation by the uppercase letters A through Z; subclades are then designated by numbers, and sub-subclades by lowercase letters and further numbering. Hundreds of thousands of partial mtDNA sequences have been generated that address human maternal lineage evolution, and these have been used to generate haplotype data for tens of thousands of humans. A phylogenetic tree of the haplotypes is presented in figure 7.4, and shows the basal position of the L group haplotypes from African populations. There are two major haplogroup branches from this basal part of the tree—the M and N haplogroups. A major haplogroup (R) is nested within the N haplogroup. Note that no single geographic region of the globe accommodates its own monophyletic group. The pattern in this tree simply represents the clonal behavior of mtDNA, and the fact that human females have moved across the globe in a pattern that can be interpreted as seen in the diagram.

This way of interpreting things has led to a very clear and well-accepted picture of our ancestors' movements around the planet and, happily, is in substantial agreement with the paleontological and archaeological evidence. Specifically, the human tree is rooted in Africa, which also shows more genetic diversity than the other continents. Most of the genetic variation outside Africa is a subset of the variation found within Africa. mtDNA genetic diversity also decreases with increasing distance from Africa. All of this points to an origin of our species somewhere in Africa, and subsequent diversification both outside and within that continent. From the data available, we can readily say that the first migration of mtDNA haplotypes out of Africa went all the way to Australia, and occurred before fifty thousand years ago. Not too much later, a second migration out of Africa occurred, resulting in colonization of Eastern Asia and Asia Minor by *Homo sapiens* at

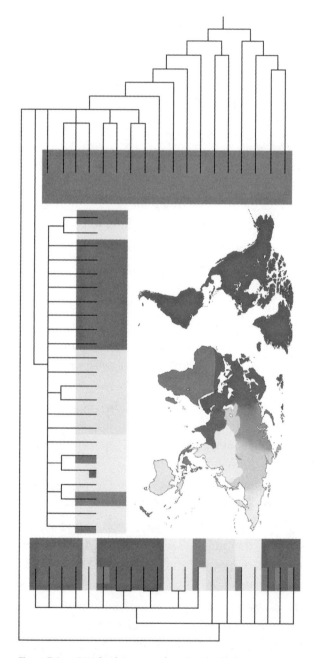

Figure 7.4 mtDNA haplotype tree based on haplogroup data from around the globe. The geographic origin of the haplogroups is designated by color. We have removed the haplotype designations for clarity. Redrawn from the Human Mitochondrial DNA Haplogroup website (www.revolvy.com/topic/Human%20mitochondrial%20 DNA%20haplogroup&item_type=topic). See plate 3.

some point between fifty and forty thousand years ago (at which time the last record of the long-resident *Homo erectus* occurs in Java). This migration into Asia was followed by migration of mtDNA haplotypes from Asia into Europe at about forty thousand years ago. Interestingly, this is about the time that our extinct European relative *Homo neanderthalensis* finally disappeared. There was then a relatively long lapse of time before *Homo sapiens* crossed the Bering land bridge (or paralleled it by sea) between about seventeen and fourteen thousand years ago, ultimately colonizing the Western Hemisphere. The haplotypes of Native Americans make an interesting subject in themselves, appearing for the most part to arise from a divergence event that occurred somewhere in central Asia. Last, but not least, at about four thousand years ago there was concerted colonization of the Pacific by mtDNA haplotypes that are closely related to haplotypes from adjacent Asia. Not all of the times just quoted come solely from molecular clocks, because the fossil and archaeological records also allow for strong absolute timing of these migrations. The patterns of divergence of maternal lineages of modern humans within these major geographic regions are also fascinating, but are beyond our remit here.

Most recently, technological advances have allowed for the rapid sequencing of entire mtDNA genomes, and currently there are at least 40,000 whole mtDNA genomes in the database (MITOMAP). In addition, the databases contain over 500 ancient human DNA mitogenomes that include both *Homo neanderthalensis* and Denisovans (see chapter 9). About half of the ancient mtDNA genomes come from specimens found in Europe, and the great majority are from modern human fossils no older than ten thousand years. Many other mitochondrial genomes exist for living humans as the by-products of clinical and other studies, and the future promises a more complete picture of human maternal lineage expansion based on those whole-mtDNA genomes. We doubt that any huge surprises will emerge that might modify figure 7.4 significantly, but the precision with which we can detect past population movements from human maternal lineages will certainly increase.

In parallel with the maternal lineage analysis done using mtDNA, researchers also developed methods to examine Y chromosomal variation in human males. The Y chromosome is a good counterpart to mtDNA, because it is inherited from father to son, making the Y chromosome the perfect tool to follow paternal lineages. The first Y chromosomal haplogroup work was accomplished by Luca Cavalli-Sforza and colleagues and Mike Hammer and colleagues. This initial work pointed to extremely deep Y chromosomal lineages, and again produced a coalescent ancestral haplotype called, not surprisingly, "Y-chromosome Adam." Again, note that this does not imply an actual individual, but rather a haplogroup of the Y chromosome. The nomenclature for haplogroups on the Y chromosome is hierarchically arranged in much the same way as for mtDNA haplogroups; and indeed, Mannis van Oven and his colleagues who organized the mtDNA haplogroups have also been involved in systematizing the Y haplogroups. Their Y chromosome tree has become the standard for understanding paternal lineage divergence.

The most recent worldwide Y chromosomal analysis was accomplished by the 1000 Genomes Project, an endeavor we will discuss at length later. In 2016 this consortium, led by G. David Poznik, took the Y chromosomal sequences of 1,244 human males from 26 populations scattered across the planet; the results are shown in figure 7.5. Remarkable amounts of variation exist on the Y chromosome (more than 65,000 variants for these 1,244 individuals alone), and this variation was used to construct a time-calibrated phylogenetic tree. The topology of the Y chromosomal tree from this meta-analysis happily agreed with the one calculated by van Oven and colleagues. The original observation that African Y chromosomal lineages are basal in the tree is borne out by this more complete analysis. One interesting novel result is that the divergence of paternal Y chromosomal lineages seems to indicate bursts of extreme expansion in the numbers of males, at times coincident with documented migrations and technological innovations.

One important thing to remember is that mtDNA and Y chromosome DNA are independent of one another. Some time ago we illustrated this with Charles Darwin's pedigree, first published at the Third International Congress of Eugenics in the 1930s. Since that time, Darwin's lineage has expanded a bit, and several individuals have been added to the pedigree. When we traced Darwin's maternal lineage (his mitochondrial history) and his paternal lineage (his Y chromosomal history), we found that the two showed very different patterns (figure 7.6). And this goes equally for the histories of whole populations containing particular haplotypes.

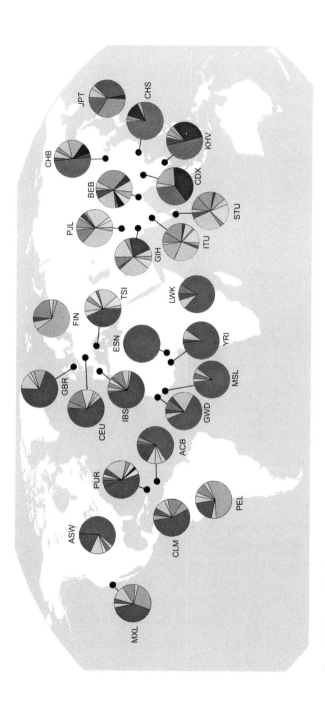

Figure 7.5 Geographic distribution of Y chromosomal haplogroup variation. Note that only one population is fixed for a Y chromosomal haplogroup (in Africa) and that all other localities indicate some degree of similarity to other localities on the globe. Redrawn from Poznik et al. (2016). See plate 4.

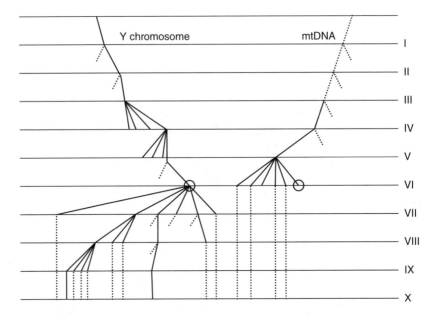

Figure 7.6 Tracings of Charles Darwin's Y chromosome (*left*) and mtDNA (*right*) from the "Pedigree of the Darwin-Wedgewood-Galton Family" first presented at the infamous Third International Eugenics Conference in 1932. Solid lines represent actual ancestor-descendant events, while vertical dotted lines are "hypothetical" descendants traced out to facilitate drawing of the trees. Slanted dotted lines represent potential branch points, to demonstrate generational branches. The black circles represent Darwin's position in the network. Note that there are no descendants coming from Darwin (or his brother) as he (and his brother) are mitochondrial DNA dead ends and none of their sisters reproduced. Horizontal dotted lines represent the generation boundaries in Darwin's pedigree, and the Roman numerals represent generation numbers, starting with Darwin's great-great-great-grandparents in generation I and ending with the current descendants in generation 10, of which there are two male descendants with his Y chromosome still alive today. Redrawn from Tattersall and DeSalle (2012).

The progression from blood group research, to allozyme research, to mtDNA and Y chromosomal DNA sequence work is an amazing story of scientific persistence. The inferences about human population divergences that researchers have made from these approaches have been rather like peeling an onion, revealing layer after layer of complexity. We will now explore even deeper layers by delving into the variation in the nuclear genome. This contains a far greater percentage of our genetic heritage, and it harbors even more surprises.

8

•　•　•　•

The Other 99 Percent of the Genome

So far, we have discussed only a very small part of our genome. The mtDNA genome is only sixteen thousand base pairs in length, with a miniscule thirteen protein-coding genes; and while the Y chromosome is much larger, clocking in at fifty million base pairs, it still only has a paltry seventy or so genes. The Y chromosome is known as a "sex chromosome," because biological sex in humans (and most mammals) is determined by its presence or absence. Your mother has two X chromosomes, and will always pass an X along to you. Your father has one X and one Y chromosome, so the X you get from your mother will be paired with either an X or a Y from your father; and if you happen to get a Y, your biological sex (that is, what genitalia you develop) will be male. Sexual identity is another issue altogether; but whether you have male or female genitalia is controlled by the presence or absence of the Y chromosome.

The approximately fifty million base pairs in a Y chromosome represent a rather unimpressive 1.6 percent of all the three billion bases in a human genome. The X chromosome, the other "sex-limited genetic element," is rather bigger; but the vast bulk of our genome resides in the other 22 pairs of chromosomes, known as "autosomes." That other 98.4 percent of your DNA harbors an immense amount of variation. What is more, while the two DNA markers we have just discussed are clonal, so that the only variation that is introduced to the next generation is caused by mutation, this other DNA experiences what is known as "recombination," a process that jumbles the alleles that a mother and father pass on to their off-spring. This occurs because the maternal and paternal DNA exchange information, and thereby create new genomic variation without muta-tion. Via this process the autosomal chromosomes become mosaics of the chromosomes inherited from their ancestors, something that poses a major problem when one tries to analyze the history of organisms—although it certainly doesn't mean that figuring out the histories of the chromosomes and their mosaic parts is impossible. What it does mean is that each little chunk of DNA on your chromosomes has a unique history. It is as if thousands and thousands of different histories are floating around in a single genome.

Sequencing the first human genomes was a very labor-intensive and expensive process. The best estimate for the cost of sequencing the first couple of human genomes is in the billions of dollars. This high cost was due to the technology used, a brilliant but cumbersome process called Sanger sequencing (named after Fred Sanger, the double Nobel laureate who invented the technology). It required the target DNA to be broken into millions and millions of small fragments, each between five hundred and one thousand bases long—with one thousand being the longest fragment the technique could deal with. With three billion bases in the genome to sequence, a few at a time, this was obviously very time-consuming. But tech-niques were developed early in this century to parallelize the sequencing steps and allow hundreds of thousands of small fragments to be sequenced simultaneously. This approach allowed the genome of James Watson (codis-coverer of the structure of DNA) to be sequenced at a cost of a million dollars. Subsequent improvements of the technology now permit billions of small fragments of DNA to be sequenced at a time. In addition, techniques have been developed in which breaking the target DNA into small frag-ments is not necessary. These are known collectively as "next-generation sequencing" (NGS), and they have brought down the cost of sequencing a full human genome to about $1,000.

Since most of the bases in our three billion-base genomes are identical from one human to the next, "targeted sequencing" was an obvious way to expedite assays of individual human genomes for variation. This approach focuses exclusively on those positions in the genome that are variable and

Figure 8.1 Hypothetical sequences from six individuals in two populations. The eighth position (in red) on the left is fixed and differs in the two populations and would be considered an SNP ripe for further analysis. The last two positions from the right (in black) differ in only one individual each and more than likely are caused by sequencing errors or other anomalies. See plate 5.

thus informative for human genome projects. It is based on the identification of positions in the genome that are polymorphic over a wide range of humans. Figure 8.1 shows a hypothetical set of sequences of the same small stretch of the human genome. Three positions in the sequence are polymorphic, and only one would be targeted as a single-nucleotide polymorphism (SNP). This position would be a good one for targeted sequencing or placement on a sequencing array.

For targeted sequencing, probes homologous to the sequence of interest are synthesized and hybridized to DNA isolated from the individual being studied. A molecule called biotin is attached to each. There are thousands of such probes, homologous to all the regions of the genome that have SNPs and thus are of interest to researchers. Small magnetic beads bearing molecules that will bind biotin are then mixed with the DNA, and any piece of double-stranded DNA that has biotin on it will bind to one of them. A magnet is then used to separate all the beaded molecules from the rest of the mixture, and all the pieces of DNA that do not contain SNPs of interest are washed away. The beads are then removed, and the DNA is sequenced using the NGS techniques. An alternative "array" approach uses a different technology but also detects targeted SNPs. The targeted sequencing approach is perhaps the more accurate, because it allows for replication sequencing at about 100× coverage (where "coverage" refers to the number of data points for a single SNP).

Typically, several hundred thousand SNPs can be assayed by these methods. The panels are commercially available, and some are proprietary. For instance, Affymetrix offers the "Affymetrix Axiom© Human Origins Array Plate," designed specifically for population genetic analysis of modern human genomes and some archaic human ones as well. This plate can analyze more than 625,000 SNPs. Illumina, another leading sequencing company, offers TruSight One, a targeted sequencing kit that assays nearly five thousand genes with extremely high accuracy. On the proprietary consumer side, the National Geographic Society Genographic Project and the private company 23andMe offer sequencing plates that produce data for three hundred thousand and nearly six hundred thousand SNPs, respectively.

How do researchers establish which SNPs to target, both for sequencing and array resequencing? Large panels of standardized human genomes are sampled, and this is where picking your populations wisely comes in. Two of the first panels developed were called Hap Map and the Human Genome Diversity Project (HGDP). HapMap was established to create a user-friendly database of genomic information and was eventually subsumed into the 1000 Genomes Project. The 1000 Genomes panel includes more than 1,000 individuals from 26 different populations; while 3,501 samples are listed in the 1000 Genomes Project database, not all have been sequenced. Table 8.1 shows the different geographic regions that make up the sample.

In contrast, the HGDP examines about twice as many (fifty-one) different populations, but with an average of about five individuals per population. The geographic distribution of the samples in this population panel is thus more extensive than that of the 1000 Genomes Project, but has fewer individuals per population. Still, both large-scale diversity studies allowed researchers to focus on SNPs that would tell them something about the differentiation of human populations on the planet, using a process called "ascertainment."

TABLE 8.1
Makeup of the 1000 Genomes Project*

Population	N
Yoruba	186
Utah, United States	183
Gambia	180
Nigeria	173
Shanghai, China	171
Spain	162
Pakistan	158
Puerto Rico	150
Colombia	148
Bangladesh	144
Peru	130
Sierra Leone	128
Sri Lanka	128
Vietnam	124
Barbados	123
India	118
Kenya	116
Houston, United States	113
Southwestern United States	112
Tuscany, Italy	112
Hunan, China	109
Beijing, China	108
Los Angeles, United States	107
Great Britain	107
Tokyo, Japan	105
Finland	105

*More appropriately, the *2000* Genomes Project.

The ascertainment procedure starts with a reference sequence from a fully sequenced and well-characterized human genome. To obtain a panel of SNPs for analysis, other carefully chosen individuals are sequenced at what is called "low coverage," which uses just enough sequencing to obtain overlapping data at a given level. This overlap is required, because at least two reads of the data are required to assess whether the SNP is variable. The choice of individuals for the low-coverage sequencing depends on the specific study; but if one is looking at the ancestry of human lineages, one's choice will be based on geographic diversity or diversity of ancestry. Researchers using this approach assume that the individuals of choice have well-defined ancestries—an assumption we will address shortly.

The sequences from the reference and new individuals (also called the "ascertainment group") are then matched, and a set of predetermined rules is used to ascertain the SNPs that the researcher thinks will be useful in further studies. Figure 8.2 shows the process of ascertainment, based on the requirement that at least 2× coverage (at least two copies of the potential SNP must exist in the data) is needed to ascertain a SNP. Potential SNPs based on the reference sequence are shaded in the reference sequence. The sequence variant on the far left is not ascertained, because only one copy of the fragment is found in the data, violating the 2× criterion. The potential SNP fragment on the far right is also not ascertained, because it is not variable in the two new sequence reads. The two potential SNPs in the middle (downward arrows) are ascertained, because they are variable in the new reads.

Table 8.2 shows the ascertainment process for the chip designed for human origins studies. It is based on genome sequences from 1.1× to 4.4× coverage from twelve living humans, and sequences from a fossil human from Denisova Cave in Siberia. Two important things are evident from the table. First, the deeper the sequencing (column 2 in table 8.2), the more potential SNPs are discovered. This occurs because the greater the sequence coverage, the more likely one is to see variation at a given position. The second trend is the decrease in potential SNPs as the process proceeds. The reduction in validated SNPs relative to candidate SNPs by phase 2 validation is in most cases by a factor of three. Overall, the reduction of SNPs is from 1.81 million to 0.54 million. At the end of the day, 542,399 unique SNPs are placed on the sequencing chip to assay large numbers of modern individuals or an ancient human DNA sample.

Figure 8.2 SNP ascertainment. The reference sequence is shown at the top. Three areas where sequences match the reference sequence are shown. The fragments on the left and middle have one potential SNP each, while the sequence on the right has two potential SNPs. The upward arrows indicate SNPs that are *not* ascertained, while downward arrows indicate SNPs that *are* ascertained and used in further research. See plate 6.

TABLE 8.2
Ascertainment of SNPs for Human Origins Studies

Ascertainment panel	Sequencing depth	Candidate SNPs	Phase 2 valid
1. French	4.4	333,492	111,970
2. Han	3.8	281,819	78,253
3. Papuan1	3.6	312,941	48,531
4. San	5.9	548,189	163,313
5. Yoruba	4.3	412,685	124,115
6. Mbuti	1.2	39,178	12,162
7. Karitiana	1.1	12,449	2,635
8. Sardinian	1.3	40,826	12,922
9. Melanesian	1.5	51,237	14,988
10. Cambodian	1.7	53,542	16,987
11. Mongolian	1.4	35,087	10,757
12. Papuan2	1.4	40,996	12,117
13. Denisova-San	*	418,841	151,435
Unique SNPs total		181,2990	**542,399**

* Not given.
Source: From Patterson et al. 2012, supplemental table 3.

Ascertainment is tricky business. What it does is set a baseline for variation studies in human genetics. If you do it incorrectly, or in a biased fashion, problems can arise; and in 2013 Joseph Lachance and Sarah Tishkoff suggested that ascertainment problems can cause genotyping arrays to contain biased sets of pre-ascertained SNPs. Emily McTavish and David Hillis have studied the problem of ascertainment bias in animal populations, finding that two types of bias may result from ascertainment problems. The first is what is known as minor allele frequency bias. This bias can result in overrepresentation of SNPs that have high minor allele frequencies and in underrepresentation of SNPs that have low minor allele frequencies. This kind of bias will influence which SNPs are put onto a chip for analysis, and entire categories of SNPs that might influence the analysis are excluded for this reason. The second kind of bias concerns the number of individuals in an ascertainment group or subpopulation. This parameter will influence the lower limit of frequencies of alleles in populations, so that SNPs that exist in low frequencies are unlikely to be observed in an ascertainment group. This many impact inference from study results. For example, it has been argued that low-frequency alleles may be the result of recent mutations that are limited to specific geographic areas simply because they have not had time to move around. If the ascertainment bias is against these SNPs, then extra information on geographic clustering of alleles for the SNP would be excluded from the study. If, on the other hand, those low-frequency SNPs were biased for a panel, then more geographic clustering would be inferred than warranted.

Because of these uncertainties, the answer to any question one might ask without a fully objective set of SNPs may be biased, because while the SNPs selected for the array or the targeted data set may be chosen for reasons that are fully in line with the research question, they may not be applicable to subsequent research questions. What this means is that we need to be very careful about the actual research questions we ask and how we approach answering them. This is key when we are looking at questions about human origins and history, for it is imperative to exclude the "ascertainment bias" that results from any inadequacies in the ascertainment process.

To drive this point home, consider the establishment of what human geneticists call "ancestral informative markers" (AIMs). In human genomics, these markers are supposedly fixed SNPs that differentiate between specific populations. To us, however, AIMs are the result of an extreme ascertainment process, as described in figure 8.3.

A final cautionary tale involves a recent study on the genetics of Europeans that gained a lot of press when it was first published, particularly because of its claim that the authors could identify individuals to within several hundred kilometers of their birthplace. In 2008, John Novembre and colleagues conducted a genome-level study by developing a panel of markers for different European populations. The procedure used was complicated, and the markers they eventually used are best described as AIMs for different subpopulations of Europeans. The way in which these markers were "ascertained" is illuminating. Table 8.3 shows the "trimming" process the Novembre group used to arrive at their set of markers and demonstrates how the strategic removal of individuals can lead to a preconceived

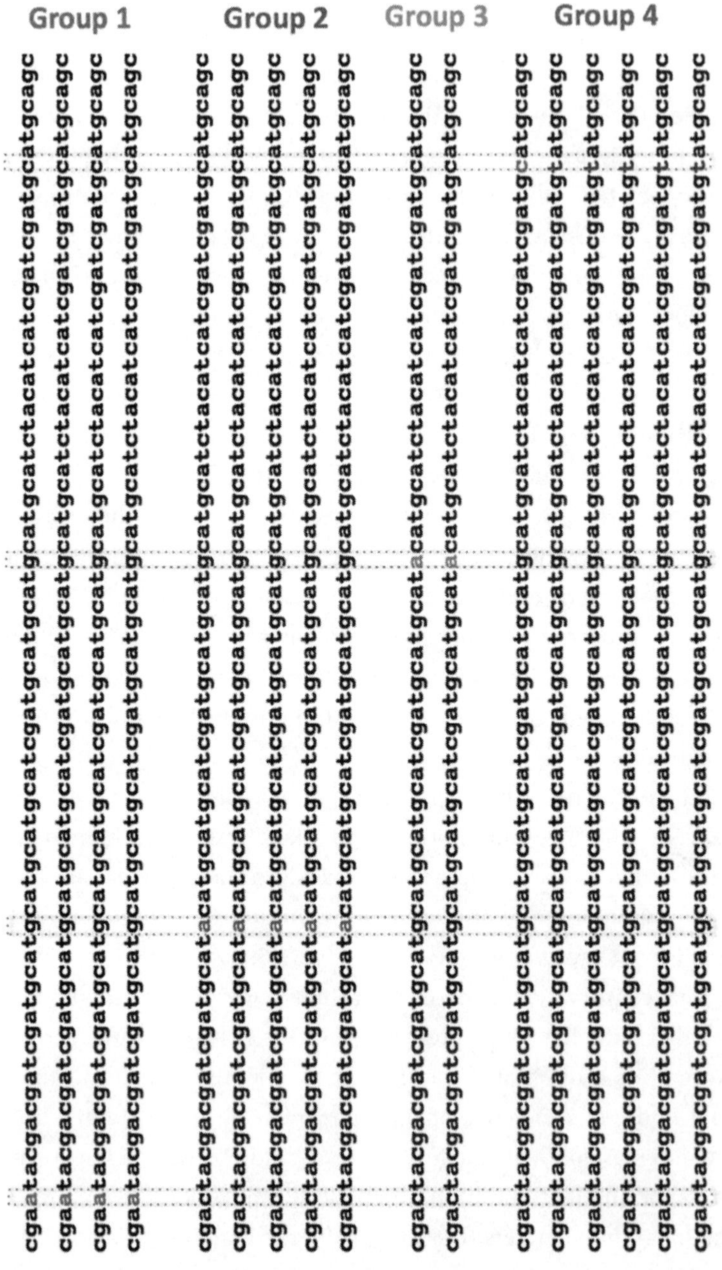

Figure 8.3 Finding AIMs in genetic data. The populations or groups are predetermined using some criterion. The sequences are then scanned for fixed and different SNPs that can be used as diagnostics. The diagnostic in group 1 is an "a" in position 4; in group 2 it is an "a" in position 20; in group 3 it is an "a" in position 40; and in group 4 it is a "t" in position 60. See plate 7.

TABLE 8.3
Exclusion Strategies and Effect on Data Set Size in an AIM Study

Sample size	Stage of analysis
3,192	Total individuals of purported European descent
2,933	After exclusion of individuals with origins outside Europe
2,409	After exclusion of individuals with mixed grandparental ancestry
2,385	After exclusion of putative related individuals
2,351	After exclusion based on preliminary run

notion of genomic differentiation among these European subpopulations. After the trimming, the researchers had removed nearly 800 out of 3,200 individuals. In other words, 25 percent of the individuals were rejected because they were "outliers." As if this were not enough, for ease of computation another one thousand or so individuals were removed to further reduce the data set, eliminating individuals with genome sequences apparently related to others in the sample.

We suggest that removing outliers in this way causes researchers to miss something incredibly important about the genetics of human populations. Sure, when you cherry-pick a data set you will obtain a neat answer that agrees well with the rules under which you did your cherry-picking. But the really interesting aspect of the Novembre group's study is actually the 25 percent of the population that was left out. Because what does an "outlier" really mean here? In the real world, outliers are most likely migrants or admixed people. The large proportion of individuals (a quarter!) who fall into this category is very significant in and of itself; and the actual proportion may in fact even be larger than indicated in this study, because many of the 1,300 or so individuals left at the end of the trimming process could be removed using similar procedures to make the results even cleaner. Indeed, it turns out that to get a geographical assignment within 800 kilometers with 99 percent accuracy, you would have to remove another six hundred or so "outliers." And this, of course, would place about half of the population in the interesting category of people who have migrated or admixed. To us, this aspect of the variation observed is by far the more interesting part and gives us a much better description of reality.

Finally, it seems highly probable that, under current demographic trends, the structure that Novembre and his colleagues were striving to document in human populations is fated to erode even further in the next few centuries (should our species make it that long). Consequently, we profoundly doubt the utility of structured analysis as a way of usefully describing human populations over the long term. We will return to these approaches in detail when we describe the pitfalls of clustering.

9

• • • •

ABBA/BABA and the Genomes of Our Ancient Relatives

Two decades ago the idea that human geneticists could ever generate the genomes of individuals belonging to extinct species of our genus *Homo* was unthinkable for many technical reasons. Since then, however, the technology has moved along so quickly that obtaining the genomes of extinct or long-dead individuals has by now become almost commonplace. The advances that have allowed this amazing technical breakthrough have involved both laboratory refinements and "informatics" tricks that depend on computer wizardry.

As an example of what has been achieved, imagine that a Neanderthal passed away some forty thousand years ago, at the impressive old age of forty. His remains would have decayed via both microbial and molecular interactions, a set of processes on which his environment of interment would have had some impact. When we and other organisms are alive, we are packed with microbes. Most of them are harmless or even beneficial to us. We need to stave off the most nefarious of them, but for the most part we live in relative harmony with these tiny, single-celled cohabitants, our immune system constantly working to keep the worst of them from getting out of control. But when we die our immune system stops working, and our remains become a haven for microbes that were kept at bay or only infested us in small numbers when we were alive. So it was for Neanderthals, too.

Once those destructive microbes had taken hold of his body, our deceased Neanderthal would have begun to decompose via those microbial and molecular processes. As the nutrients in his corpse were used up, the microbes themselves would have started to die, leaving their own DNA hanging around the fossilizing tissues. If the Neanderthal had died in a moist, humid resting place, so much for the DNA of Neanderthal and microbes alike: it has been estimated that intact DNA can only exist in the presence of significant amounts of water for something like ten thousand years, at which point it will have been entirely fragmented into single bases and be useless for DNA sequencing. But if, on the other hand, our Neanderthal had happened to pass away in a dry, cool cave, his DNA would have broken down much more slowly. His fossilizing corpse would still have had loads of microbial DNA bound up in it; but at least all his own DNA would have stayed relatively intact for longer. And although his fragmented DNA would nonetheless be in pieces not much more than a hundred bases long by the time a few tens of thousands of years had passed, with the right algorithms and with intensive computation, it might still be possible to reassemble his DNA from the fragments.

The problems of microbial contamination and the extensive fragmentation of the DNA strands are also complicated by the fact that people working with the bones leave traces of their own DNA on the fossils, both as they are excavated and as they are worked on in the lab. The resulting complications are fiendishly daunting; but one diligent and dedicated researcher has made all the difference in this field. Svante Pääbo, now at the Max Planck Institute for Evolutionary Anthropology in Leipzig, Germany, dedicated his career to working on these problems, and he and his colleagues eventually overcame all the technical difficulties using clever biochemistry and computational methods. In doing so, they established the field of ancient human genomics, in which several other laboratories are now also active.

Most of the archaic human fossils sequenced so far have been less than forty thousand years old. However, in some cases of exceptional preservation, the boundary can already be pushed. The oldest human DNA so far analyzed comes from five bones, among several thousand found in the Sima de los Huesos ("Pit of the Bones") in northern Spain, that date back to around four hundred thousand years ago. And perhaps even more remarkably, DNA has permitted the detection, from one tiny finger bone found in Siberia's Denisova Cave, of an entirely new kind of extinct hominid that could not have been recognized in any other way. Indeed, we still do not know what those extinct "Denisovans," contemporaries of both Neanderthals and the earliest European modern humans, would have looked like physically. We just know they were genomically distinctive.

The addition of ancient genomes to the fray has facilitated a whole new dimension of analysis and hypothesis testing in paleoanthropology (figure 9.1). Before ancient DNA could be easily manipulated for whole-genome sequencing, when tissue from living or recently dead individuals was required the kinds of questions that could be asked were limited. What is more, the ease with which data on clonal DNA could be generated and analyzed ensured that the most studied quadrants in figure 9.1 were the clonal ones. But recently, the upper quadrants have been the focus of a lot of research, with worldwide studies in the HGDP and 1000 Genomes Project focusing on the upper right quadrant.

Sequencing ancient mtDNA from archaic Neanderthal and Denisovan specimens first focused on sequencing what is called the HV1 region of the

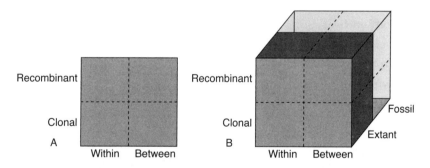

Figure 9.1 Human origins analysis space. The left diagram shows the space without ancient DNA. The y-axis determines whether the DNA comes from mt genomes or the Y chromosome (clonal) or the autosomal nuclear and X chromosome (recombinant). The x-axis determines whether the analysis is done within or between taxonomic units. Addition of the temporal framework (z-axis) makes for a widening breadth of questions to be asked. So, for instance, a population study of Taiwanese mtDNA lineages would fit in the lower left quadrant of figure A and a worldwide survey of Y-chromosome variation would fit into the lower right quadrant of figure A. A study of genomic variation of people from the British Isles would fit into the upper left quadrant, and a study of variation of chromosome 21 across European populations would fit into the upper right quadrant and so on.

mtDNA. The HV1 sequence is a non–protein coding region of the mtDNA genome that is variable enough to allow for fairly accurate placement of its variants in a phylogenetic tree. The phylogenetic tree of this clonal marker revealed several longish basal branches of Neanderthals, mostly from Central Asia, and a cluster of more recently diverged Neanderthals (circa fifty thousand years old) exclusively from Europe. The authors of this analysis suggest that the European Neanderthal populations diverged before our own *Homo sapiens* migrated into Europe.

Several Neanderthal and a few Denisovan complete mtDNA genomes have been sequenced, and used to address Neanderthal relationships and the placement of Denisovan specimens in the human family tree. They also revealed a great deal of variation among the maternal lineages of these extinct human taxa. By now, the nuclear genomes of three Neanderthal specimens and one Denisovan specimen have also been sequenced to low coverage. These studies fit into the rearward quadrants of figure 9.1, because they have been compared with a wide array of living *Homo sapiens* genomes. Figure 9.2 shows the phylogenetic tree generated from the seven full Neanderthal mtDNA genomes and two Denisovan mtDNA genomes currently available, along with several human lineages. This tree is similar in many ways to the tree generated from HV1 region sequences. In both analyses, the Denisovan sequence was used as an outgroup to root the trees. Addition of an mtDNA genome sequence from a 430,000-year-old specimen from the Sima de los Huesos (figure 9.2) suggested that this older specimen's mtDNA is more closely related to Denisovan than to Neanderthal mtDNA.

In a truly spectacular development in 2017, researchers at the Max Planck Institute, led by Pääbo, guessed that DNA from archaic humans might be recoverable from the sediment present at archaeological sites. Their reasoning stems from the idea that remains or products of organisms can complex with minerals in the sediment; and while they might not necessarily be preserved as visible fossils, their nucleic acids just might still be there. Because the sediment will include microbial DNA and that of any other animals that might also have been present (bovids, canids, felids, rodents, elephants, ursids, cervids, etc.), special sequence data analysis filters were developed to separate the human sequences from everything else. In addition, the researchers obtained DNA from

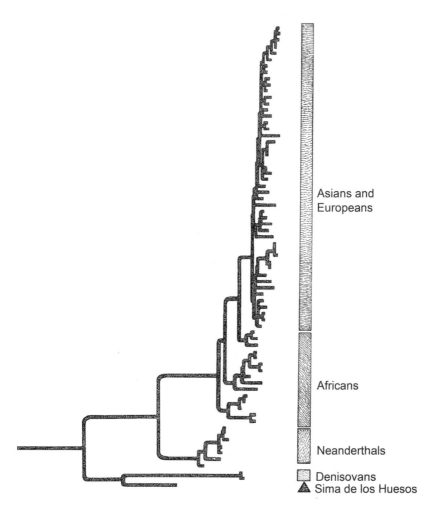

Figure 9.2 Phylogenetic tree of Sima de los Huesos, Neanderthals, *H. sapiens*, and Denisovan specimens for whole mtDNA genomes. The Bayesian approach was used to generate the tree. The Bayesian posteriors at the node defining Neanderthals is 1.0, which suggests a high degree of confidence of the existence of that node. Redrawn from Skoglund et al. (2014).

different sedimentary layers, which allowed them to determine which humans were present during different time frames. Their work has recovered a familiar phylogenetic tree from these samples, but it shows that the Denisova Cave had several Neanderthal residents in addition to the Denisovans (figure 9.3). The study also adds to the range of Neanderthal variation available.

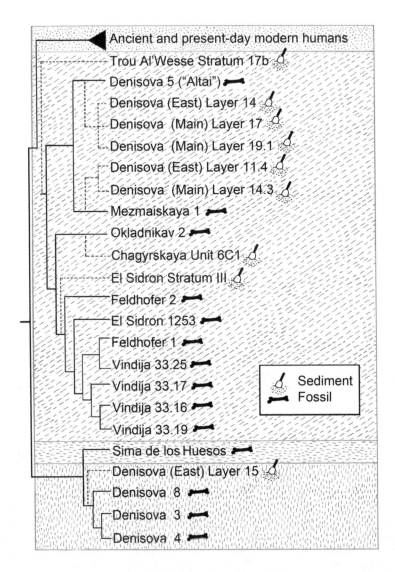

Figure 9.3 Phylogenetic tree of bone-derived complete mtDNA sequences from Sima de los Huesos, Neanderthals, *H. sapiens*, and Denisovans along with sediment-derived mtDNA genomes from Neanderthals and Denisovans. In this tree, the locations of the archaic hominins are used as terminals. Redrawn from Slon et al. (2017).

The mtDNA results seem to show that, among the taxa analyzed, *Homo neanderthalensis* and *Homo sapiens* are each other's closest relative, with Denisovans and Sima de los Huesos populations showing closer affinity to each other than to either *H. sapiens* or Neanderthals. But what about the rest of the genome? As we have already pointed out, sequencing nuclear genomes is much more difficult than sequencing mtDNA; and while hundreds of ancient mtDNA genomes and tens of thousands of living *H. sapiens* mtDNA genomes have been sequenced, it seemed for a while that large numbers of fossil nuclear genomes would be unattainable due to the expense and labor involved. But again, Pääbo and other researchers have developed techniques to more easily secure archaic human and ancient human genomes. The added temporal dimension, and the ability to include nuclear genomes, opens up a wide range of novel and essential questions about human origins and ancestry.

The biologists Montgomery Slatkin and Fernando Racimo pointed out in a recent review of ancient human genomes that nearly one hundred such "paleogenomes" have been sequenced in the past six years, most of them completed in 2016. And it is hardly an exaggeration to claim, as Slatkin and Racimo did, that a "paleogenomic revolution" has occurred in a very short span of time. Figure 9.4 shows the rapidity with which this field has developed. One of the major developments in this timeline is the completion of the 1000 Genomes Project phases 1 and 3, because the inclusion of the 1000 Genomes Project database, made up of geographically diverse living humans, is essential to interpreting what we see in the archaic individuals. And interestingly, in this context we see a very different picture from what we saw from mtDNA alone. For the nuclear genomes indicate that Neanderthals and Denisovans are each other's closest relatives, while *Homo sapiens* is the "odd man out." In addition, when the nuclear DNA of two Sima de los Huesos specimens is added, these humans no longer appear most closely related to Denisovans, but rather are sisters to Neanderthals (figure 9.5).

While the inferences concerning relatedness among Neanderthals, Sima de los Huesos, Denisova, and our own species are certainly intrinsically interesting, perhaps the most striking and controversial result of the sequencing of these archaic humans comes from the light they shed on

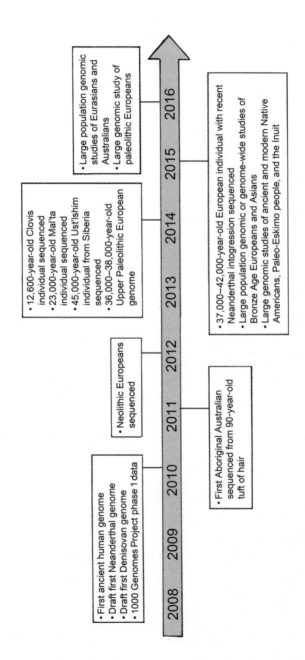

Figure 9.4 Timeline showing the development of whole-genome sequencing for ancient (fossilized modern *H. sapiens*) and archaic humans (Neanderthals and Denisovans). From Nielsen et al. (2017).

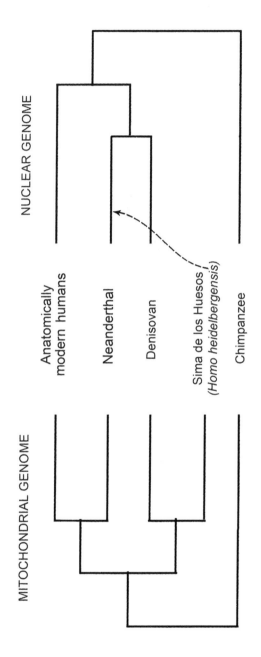

Figure 9.5 MtDNA genome phylogenetic tree (*left*) and nuclear genome tree (*right*). The trees are from Ermini (2014), but the Sima de los Huesos data were generated in 2016; hence the arrow indicates the position of the Sima de los Huesos lineage.

MITOCHONDRIAL GENOME

NUCLEAR GENOME

Anatomically
modern humans

Neanderthal

Denisovan

Sima de los Huesos
(*Homo heidelbergensis*)

Chimpanzee

the ancestral makeup of our own genomes. To do this, researchers developed what they call the ABBA/BABA test (figure 9.6). The test requires an outgroup reference sequence. Any sequence outside of the Neanderthal/ Denisova/Sima de los Huesos group will do the trick, and researchers settled on the chimpanzee lineage because it is the closest living relative for which we have genome sequences. The test is accomplished by comparing two modern human lineages (in figure 9.6 these two lineages are Yoruba and French). Finally, an archaic genome is also included (in figure 9.6 the archaic genome is from Neanderthal). The "real" history for any SNP in the

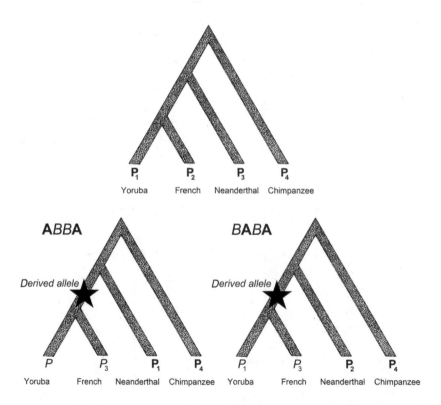

Figure 9.6 The ABBA/BABA test. The top tree is the "real" or expected tree. The bottom, left tree is the ABBA tree (French with Neanderthal), and the right tree is the BABA tree (Yoruba with Neanderthal). Individuals are numbered as P1 (Yoruba), P2 (French), P3 (Neanderthal), and P4 (chimpanzee). The agreement of the SNP with the tree is accomplished by examining the state at the node with the star labeled "derived state."

genome is displayed at the top of the figure. In this diagram, the two modern human groups should be sister terminals. However, there will always be some SNPs that do not reveal this history. Remember from chapter 3 that there are three possible trees when we have three ingroups and one outgroup. So there are two other possibilities—one in which the first modern human group is sister to Neanderthals, and one in which the second modern human group is sister to Neanderthals (French/Neanderthal vs. Yoruba/Neanderthal).

One simply sifts through the sequence data and assesses for each SNP whether it supports the "real" tree—and, if not, which of the other two trees the SNP supports. The test then compares the tallies for the three trees. The number used to summarize the analysis is called the D statistic, and is given by the simple ratio of the number of ABBA SNPs minus the number of BABA SNPs, divided by the number of ABBA SNPs plus the number of BABA SNPs. If the D statistic is positive, it means that there are more ABBA SNPs than BABA SNPs, and hence that more Neanderthal SNPs resemble Yoruba SNPs. On the other hand, if the D statistic is negative, there are more BABA SNPs, meaning that more SNPs support a Neanderthal-French relationship. If the evolution of all the SNPs is regular, the D statistic should be 0.0. But if there is something odd occurring, a positive or negative D statistic is obtained, and the magnitude of the D number tells us how much hanky-panky is going on. Table 9.1 shows the results of applying the ABBA/BABA test to several modern human individuals. Note that in the D column of the table the only values that are significant are comparisons of non-African with African, and that the values are between $D = -3.8$ and $D = -5.3$. These results indicate that there is an excess of Neanderthal-like sequences in non-African genomes.

But what kind of hanky-panky could cause a nonzero D statistic for a comparison? Remember that the nuclear genome undergoes recombination during sexual reproduction. One mechanism might be interbreeding between the Neanderthal and *Homo sapiens* lineages. This would not be unexpected, since very closely related species are often observed to interbreed if they have the opportunity—even though, if they truly are different species (as seems to have been the case with Neanderthals and ourselves), this behavior will not lead to reintegration of the two lineages.

TABLE 9.1

ABBA/BABA Results from the First Neanderthal Genome Sequence Paper

H1	H2	H3	Number ABBA	Number BABA	D	Z score	Interpretation
San	Yoruba	Neanderthal	99,515	99,778	−0.1 ± 0.3	−0.4*	Neanderthal equally close to Africans
French	Han	Neanderthal	74,477	73,089	0.9 ± 0.5	1.7*	Neanderthal equally close to non-Africans
French	Papuan	Neanderthal	70,094	70,093	0.0 ± 0.5	0.0*	Neanderthal equally close to non-Africans
Han	Papuan	Neanderthal	67,022	68,260	−0.9 ± 0.6	−1.4*	Neanderthal equally close to non-Africans
French	San	Neanderthal	95,347	103,612	−4.2 ± 0.5	−9.3	Neanderthal gene flow with non-African
French	Yoruba	Neanderthal	84,025	92,006	−4.6 ± 0.4	−10.5	Neanderthal gene flow with non-African
Han	San	Neanderthal	94,029	103,590	−4.8 ± 0.5	−9.9	Neanderthal gene flow with non-African
Han	Yoruba	Neanderthal	82,575	91,872	−5.3 ± 0.5	−10.5	Neanderthal gene flow with non-African
Papuan	San	Neanderthal	90,059	97,088	−3.8 ± 0.5	−7.0	Neanderthal gene flow with non-African
Papuan	Yoruba	Neanderthal	79,529	85,570	−4.2 ± 0.6	−7.5	Neanderthal gene flow with non-African

* Z scores are significant.
Source: From Green et al. (2010).

Still, if Neanderthal individuals had mated with members of our species during their brief time of geographic overlap, then part of the Neanderthal genome might have recombined into *H. sapiens* genomes. In which case, we would expect chunks of the *H. sapiens* genome to resemble their counterparts in Neanderthals, making the *D* statistic nonzero. Alternatively, simple convergence of DNA sequences might have occurred because of independent mutations in the two lineages, leading again to a nonzero *D* statistic. This is also nothing unusual, though it requires a completely different mechanism. And there is a third way in which the *D* statistic might be nonzero: the SNPs might have sorted before speciation, something that also happens more frequently than you might think. All these possibilities are perfectly okay as scenarios to explain a nonzero *D* result, especially when you consider what we have said about genes in the genome often having different histories. But one helpful thing to consider in choosing among them is that is that chunks of DNA will be convergent in the hybridization scenario, whereas in the other two scenarios, the convergent SNPs will be randomly dispersed.

The first study using this approach estimated that up to 4 percent of the genomes of non-African individuals could be said to have chunks of Neanderthal DNA embedded in their genomes. Using the same approach, but using Denisovan instead of Neanderthal SNPs, and comparing them to the genomes of Oceanian people, the researchers later suggested that up to 6 percent of the Oceanian genome is made up of introgressed Denisovan genomic DNA. A further study, using more Southeast Asian, Oceanian, and Australian modern human genomes, claimed to pin down the general location of interbreeding to Southeast Asia. This last inference also suggests that Denisovans occupied a very broad range of habitats, stretching all the way from the frigid Russian steppes to the steamy tropics. Only one other human species—*Homo sapiens*—has ever occupied such a wide range of habitats.

The controversy over whether biological mixing is the cause of the observed ABBA/BABA patterns is a genuine one, and several studies have challenged the validity of blaming sexual hanky-panky for these patterns. For instance, Anders Eriksson and Andrea Manica have tweaked the population structure models used in ABBA/BABA measures to reach the conclusion that there is no statistical difference between the probabilities of the sorting and interbreeding scenarios. In a similar vein, William Amos (2016) has recently adjusted the assumption that mutation rate needs to be constant to show that slightly altered mutation rates in the derived *Homo sapiens* populations could produce the divergent ABBA/BABA patterns with respect to Neanderthal comparisons with non-African lineages.

The estimate of the quantity of Neanderthal introgression seems recently to have dwindled to about 2 percent, while the claim that there was no introgression of Neanderthal DNA in African populations has recently run into challenges. In combination with scientists' capacity to alter the inferences by tweaking the assumptions of the models used, these findings suggest that the dramatic scenario involving hybridization between the morphologically very highly divergent *H. neanderthalensis* and *H. sapiens* continues to be a moving target. On the other hand, based on current information, the Denisovan introgression story seems to be on firmer ground.

Still, the story of the Neanderthals, Denisovans, and modern humans does carry a very clear moral. Whatever it was that actually happened, all three lineages evidently maintained their distinctive identities, and they certainly experienced independent evolutionary fates. And whether—or not—members of modern *Homo sapiens* boast any "Neanderthal genes," they remain very much themselves, at most only marginally affected by any ancient hanky-panky. This picture contrasts starkly with what we see today. For even though lineages have clearly diverged within our unprecedentedly geographically widespread species, those lineages invariably show an irresistible tendency to reintegrate when they have the opportunity.

10

• • • •

Human Migration
and Neolithic Genomes

The genomics of extinct humans such as Neanderthals and Denisovans help us understand some very important facets of own species. The interbreeding scenario aside, comparing those paleogenomes (both mtDNA and nuclear) with our own sheds light on three major aspects of *Homo sapiens* evolution.

First, it emerges that all living members of our species are very closely related, and distinct from our close extinct relatives. In addition to the genome sequences of members of those extinct species (Denisovan and Neanderthal), and the large-scale genome sequencing studies of large numbers of living humans, many genomes of fossilized *Homo sapiens* have also been sequenced. Montgomery Slatkin and Fernando Racimo summarized the state of the paleogenomic revolution and counted nearly 150 ancient genomes that have been sequenced since 2008 through either whole-genome sequencing or high-density chip sequencing (figure 10.1).

The figure demonstrates that Europe has been the major focus of ancient genome sequencing and that there is a preponderance of genomes from the one thousand to ten thousand years ago age range. For this reason, the inferences made from paleogenomes are most complete for Europe during the Neolithic (New Stone Age) period. There is a dearth of genomes from almost all other regions of the globe, especially Africa and East Asia. In Africa this is easily explained, as the conditions for preservation of DNA

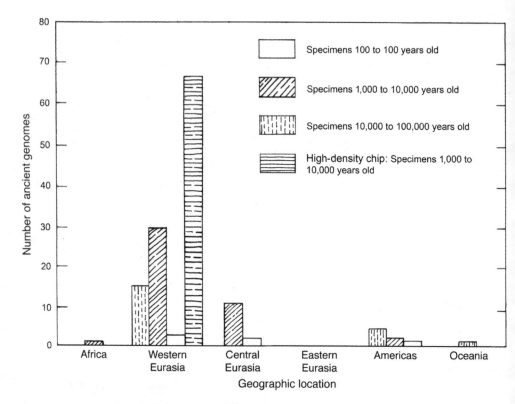

Figure 10.1 Chart showing the distribution of ancient genomes sequenced from the specified geographic areas. The number of genomes is shown on the *y*-axis, and the geographic location is on the *x*-axis. From Slatkin and Racimo (2016).

in fossils are not easily met in the sub-Saharan region. And, aside from inhospitable environmental circumstances, Africa and Asia are relatively poorly known, paleoanthropologically. Nevertheless, Rasmus Nielsen and colleagues have recently summarized what we have learned from paleogenomics about how our ancestors moved out of Africa and across the planet.

Since the early 2000s it has been established that what were effectively anatomically modern humans were already living in the eastern African region between about 200,000 and 160,000 years ago. There are claims for "early modern *Homo sapiens*" in northern Africa before this, but the fossils involved have archaic features, so we can currently put a lower limit of about two hundred thousand years ago for the appearance of our modern species in Africa. Happily, this conclusion agrees pretty well with estimates derived from both genomic and mtDNA data. The genomic information also suggests that our species had a pretty complex history in Africa before spreading out from the continent. In fact, the genomic data indicate that three distinct major African lineages diverged from one another early on, to produce a greatly subdivided population structure. These divergences are of deep (ancient) origin: they are, indeed, the deepest of any of the splits we find encoded in the genomes of modern human populations. A highly detailed study using some three thousand individuals of African ancestry further subdivided these three major lineages into a total of fourteen lineages.

The deepest living human lineage diverged from the ancestral African population between 160,000 and 110,000 years ago, a good 40,000 to 90,000 years before our species started migrating out of Africa. According to Nielsen and colleagues, there were several subsequent events that then shaped the modern structure of human populations in Africa. The biggest migration event within Africa, and one that produced a lot of recontact with other lineages, was a major movement of Bantu-speaking populations from the highlands of Central Africa to other areas of sub-Saharan Africa about four thousand years ago. Two earlier migration events, one at seven thousand years ago and another at five thousand years ago, already had discernible impacts on the structure of African populations. The first dispersal of fully modern people out of Africa had occurred long before these, at around sixty thousand years ago. And while that event probably had negligible impact on the populations remaining in Africa, it led to drastic alterations in the genomic structures of those that migrated.

The fossil record relevant to humanity's first steps out of Africa is a little hazy. There is some fossil evidence to suggest that anatomically modern humans had made it into the Neanderthal-occupied neighboring Levant more than one hundred thousand years ago; but these people apparently never gained a permanent foothold there. This may have been because they were still behaving in the archaic human manner, for at this period humans were not yet leaving behind archaeological evidence of behaviors related to the radically new form of symbolic reasoning that is such a striking characteristic of *Homo sapiens* today (such evidence begins to show up in Africa in the one hundred thousand to eighty thousand year time frame). A new fossil date from China suggests that anatomically modern—and potentially symbolically behaving—humans might have made it to the Far East by about eighty thousand years ago; and they were certainly in that region by sixty to forty thousand years ago. Modern humans had reached Europe by well over forty thousand years ago, and shortly thereafter had begun creating the extraordinary tradition of cave art that is the most stupendous and overwhelming early evidence we have for the flowering of the modern human creative spirit.

In 2016, three large studies were conducted using the genomes of people from 270 geographically diverse localities. These studies suggested that the patterns of population divergence of non-African populations can best be explained by an early migration event that ultimately went as far as Oceania and was followed by a further major wave of migration out of Africa. The principal impact on modern human population structures outside Africa was made by the latter wave, but echoes of the earlier migration to Oceania are also detectable in modern genomes. The details of the dispersal routes are hard to infer from current data, but it seems that the first wave traveled to the areas of Australasia and New Guinea, while the second supplied the ancestors of the current Eurasian populations.

The paleogenomic data add very importantly to what we can infer from current genomes. Three fossils, and the inferences made from the comparison of their genomes, are listed in table 10.1. The Mal'ta-Buret paleogenome suggests that twenty-five thousand years ago, a genetic structure existed wherein eastern Asian lineages were already becoming distinct from the western Eurasian ones. The Ust'-Ishim paleogenome indicates that twenty thousand years earlier there had been relatively little genetic differentiation and that the two main lineages had not yet diverged much. These two paleogenomes thus give us between forty-five thousand and twenty thousand years ago as the time range within which the major Asian lineages diverged. Adding the Kostenki 14 paleogenome narrows the range down even further, to between forty-five thousand and thirty-seven thousand years ago.

TABLE 10.1
Fossils and Genomes Weigh in on the Divergence of Asian People from Western Eurasians

Fossil	Location	Age	Observation
Mal'ta-Buret	Siberia	25,000	Genetic affiliation to both western Eurasians and Native Americans, but a weaker affiliation to East Asians and Siberians
Ust'-Ishim	West Siberia	45,000	Almost equal genetic affinity with western Eurasians, East Asians, and Aboriginal Australians
Kostenki	Russia	37,000	Close affinity to contemporary western Eurasians, but not East Asians

In the other great surge of people out of Africa and into Asia, fossil evidence pins down the first presence of *Homo sapiens* in the Oceanian region at about fifity-five thousand to forty-eight thousand years ago. Two separate migrations to the ancient continent of Sahul (Australia, New Guinea, and Tasmania, which were contiguous during the ice ages, when sea levels were lower) are inferred based on fossil, linguistic, and other data. In contrast, paleogenomic data suggest only a single migration to Sahul, followed by divergence due to isolation caused by environmental changes and sea-level fluctuations.

The specifics of the movements of people following these initial great migrations have also been worked out using data from extant human genomes. In Eurasia, two waves of migration had great impacts on population structure. About five thousand years ago, the Yamnaya people migrated both east and west from Siberia, and a thousand years later the Shintashta people migrated east. As figure 10.1 shows, we have no ancient genomes from eastern Asia, so no paleogenomic perspective is possible for this area. However, using the genomes of living *Homo sapiens*, it is still possible to decipher some of the migrations and movements of people in the region. The Oceanian migration, for example, also subsequently involved submigrations to Polynesia.

The power of the genomic approach here was demonstrated by Jianjun Liu and colleagues (Chen et al. 2009), who looked at 350,000 SNPs in over

6,000 individuals living in China. These researchers were interested in the dynamics of movement of the dominant Chinese Han populations, and their analysis shows a clear north versus south population structure, with more derived genomes in the south, and more ancestral ones in the north. Samples from cities like Beijing and Shanghai did not fit the pattern— understandably enough, given the large numbers of people from all over China who have migrated to these mega-urban centers.

In another study, David Reich and colleagues analyzed 132 individuals in 25 Indian populations, using more than 500,000 SNPs. Their analysis suggested the existence of two major ancient human lineages on the Indian subcontinent: "Ancestral North Indian" (ANI) and "Ancestral South Indian" (ASI). These lineages are very distinct from each other, or, as the authors put it, the ASI population "is as distinct from ANI and East Asians as they [ANI and Eurasians] are from each other" (Reich et al. 2009, 489). The researchers also showed that the ancestral ANI lineage is closely related genetically both to Middle Easterners and to Europeans and suggested that, because of extensive interbreeding, there may not be any individuals left in mainland India who have only ASI ancestry. One offshore exception to this general rule is the Andaman Islanders, who are ASI-related, with no ANI ancestry. Extensive cross-breeding has resulted in most Indian groups having only 40 to 70 percent ANI ancestry. Also in the southern Asian region, John C. Chambers and numerous colleagues (Kooner et al. 2011) examined a sample of over 150 South Asian individuals from Pakistan, India, Bangladesh, and Sri Lanka using low-coverage whole-genome sequencing and confirmed the basic divergence and the genomic differences of the ANI and ASI lineages.

Paleogenomics has made it much easier to analyze the population structures and migrations of people in the past; and Europe is the best studied of the continents when it comes to paleogenomes. As we write, more than one hundred Eurasian paleogenomes have been generated from a broad range of dates up to around forty-three thousand years ago, the age of the oldest *Homo sapiens* fossil in Europe. The earliest of these paleogenomes thus shed light on the first modern humans who entered Europe. Based on the genomes available, the best estimate is that there were three major waves of modern immigration into Europe that occurred at different times.

The first wave included the "Cro-Magnons," who left such an extraordinary record of Ice Age cave art and other creative expressions. The second began at the end of the Ice Age some 11,000 years ago; and the third started about 4,500 years ago.

The pre-Neolithic genetic history of Europe is complex, but it appears to have been dominated by a major wave of people who probably contributed rather little to present-day European genomes. The genomic signature of this wave is best detected in the paleogenomes of individuals from the pre-Neolithic period. The second (Neolithic) wave started eleven thousand years ago, with the introduction of agriculture, animal breeding, and a move away from the ancestral hunting-gathering lifestyle. Appropriately enough, the genomes of these early European agriculturalists came from the Fertile Crescent of the Near East, one of the main centers of this economic revolution. Specifically, they seem to have originated in the Anatolian region of modern Turkey. The many paleogenomes available from this time period indicate that a rapid westward movement of people out of Anatolia toward the Iberian Peninsula was complete by about seven thousand years ago and that the northern regions of Europe (Scandinavia and England) were populated by an extension of this wave around six thousand years ago. The third genomic wave saw people from the Ponto-Caspian area migrating into Europe about 4,500 years ago. As we pointed out in chapter 8, John Novembre and his colleagues suggest that genomic structure within Europe is strongly correlated with geography. The underlying genetic content came mostly from the last two waves of migration into Europe, which, as Rasmus Nielsen and colleagues put it, "can explain much of the genetic diversity found in present day Europe" (Nielsen et al. 2017, 305).

The rather tardy peopling of the Americas was also complex and was also clearly accomplished in several waves. The exact timing of the initial entry into the Americas is enigmatic, but more than likely it was accomplished between about fifteen thousand to twenty-three thousand years ago. Among living populations, Siberians are thought to be the closest to Native Americans. But there is also strong evidence that Native Americans were partially founded by populations that were closely related to the Mal'ta paleogenomic lineage (table 10.1). The ancestry of Native Americans also apparently involved input from an as yet unidentified East Asian lineage. This means that, at some time in the past, these two lineages (Mal'ta and unknown) interbred to give rise to the ancestral lineage of Native Americans. It is estimated that the interbreeding or admixture of these lineages took place about 12,500 years ago. This date is interesting, because it almost exactly coincides with the calculated splitting date of the two major lineages of Native Americans at around thirteen thousand years ago, when today's northern Native Americans appear to have diverged from those Native Americans who migrated south. Amerindians of the far north, such as the Inuit and the paleo-Eskimo, were founded by later migrations that took place three thousand and five thousand years ago, respectively. There is also some evidence of possible Polynesian migration to South America in the past three thousand years, but this latter migratory event needs more study.

Use of both the genomes of living people and the expanding library of available paleogenomes to infer ancestry and population history has uncovered several unexpected patterns of migration and admixture. And if we wanted to summarize the major lesson that has emerged from this work, we could do no better than to quote Joseph Pickrell and David Reich, two geneticists who have been at its forefront: "Empirical data have shown that the current inhabitants of a region are often poor representatives of the populations that lived there in the distant past" (Pickrell and Reich 2014). To illustrate this, Pickrell and Reich offer one of the clearer examples—the Americas. The current genomic profile of the Americas has been hugely impacted in the last five hundred years by the migration of Europeans and by the horrible practice of importing slaves that produced the movement of Africans to the Americas. These extremely recent events have shifted the modern genomic makeup of the Americas so much that it very little resembles the situation ten thousand years ago. This theme of replacement has arisen repeatedly across the planet, as shown in figure 10.2.

These observations beg for a model to explain how humans moved around the globe, and the "go-to" model for how humans moved out of Africa is what is called "serial founder effect." A founder effect occurs when a single individual, or multiple closely related individuals, migrate to a new area and "found" a new population. Since we are dealing here with

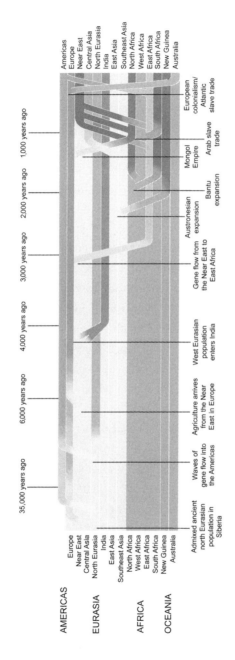

Figure 10.2 Diagram showing the complex migratory patterns of human populations in the last fifteen thousand years. Thirteen current populations are represented in the figure with different colors. The migratory patterns of these populations involve all thirteen populations and involve at least seventeen discernible events. From Pickrell and Reich (2014). See plate 8.

sexually reproducing organisms, you might query that word "single," but in fact some animals and plants can migrate long distances while fertilized. The Hawaiian Islands, for example, harbor large numbers of insect species that apparently arose as the result of the migration of a single gravid (pregnant) female. When such an event happens, if it survives, the population is inevitably propagated through inbreeding. This inbreeding is also known as a "bottleneck," because genomic variability is dramatically reduced in the new population relative to what had been present in the parent population. While possible for humans, this kind of super-extreme founder event probably never happened in our species. But our ancestors were nonetheless very thinly spread on the landscape, and bottlenecks involving small groups of closely related individuals would not have been uncommon in human history.

Serial founder events are inferred where genomic analysis reveals a decline in heterozygosity (also inferred as more and more inbreeding) as one goes from the center of origin (in our case, Africa) to increasingly remote regions of the globe. This appears to apply to *Homo sapiens*, and it makes interpretation of our genomes relative to our dispersal histories relatively simple, because as Pickrell and Reich point out, discovering where our species originated is largely a matter of finding where on the planet genetic diversity is the highest. The serial founder effect has also been used to interpret linguistic and physical anthropological data; but as with any model, the more generalized it is, the less realistic it probably is in practice. And there are other models available to explain what we see.

Specifically, two other models also show a smooth gradual loss of heterozygosity with increasing distance from Africa. The first involves at least two founder events, producing severe bottlenecks along with large amounts of migration between populations that resulted in admixture. The second model does not require bottlenecks, but invokes admixture with another species (Neanderthals, Denisovans) early in the penetration of *Homo sapiens* beyond Africa. This latter model would also require a lot of recent admixture; but since there is some evidence for both Neanderthal and/or Denisovan admixture (chapters 8 and 9) and recent population admixture (figure 10.2), this model also might also have its appeal. And that is not the limit, for there are other possible versions of our history that incorporate aspects of all three models. Still, regardless of which model is correct, or more accurate, the lesson is the same: our current genomic variability is the product of a past that was rich with interbreeding and admixture. The degree of ancient admixture remains debatable, of course; but the degree to which it occurs today is unarguable.

11

• • • •

Gene Genealogies and Species Trees

Making either/or decisions about things is a hallmark of organisms on our planet. Some scientists believe that this dichotomous way organisms have of making important decisions is a result of natural selection. If you do not categorize something as dangerous, you get eaten; if you do not categorize something as food, you starve; if you do not categorize something as a potential mate, you do not reproduce. While this is clearly an oversimplification, we do have to admit that our human view of the world is a very dichotomous one. Understanding the structure of populations is no exception; and indeed, as we saw in chapter 3, one of the classic ways of thinking about the way organisms are organized on our planet is through the branching diagrams we call phylogenetic trees. When you make any decision such as "if a group of things isn't this, then it has to be that," you are simply creating a dichotomous branch in a phylogenetic tree.

At the species level this procedure is called "phylogeography." Among other things, phylogeographic studies often test the hypothesis that individuals from different geographic regions, or individuals grouped according to some other criterion, have a common ancestor to the exclusion of all other individuals being considered. This phenomenon is called "reciprocal monophyly," and the assumption is that the reciprocally monophyletic groups are separated in some biologically meaningful way. On the face of it, this approach is very objective and operational; all you need is a lot of information from large numbers of individuals in the putative groups you want to test. The tree generated from the data then supplies the proof of the pudding. But nothing is ever that easy.

The first DNA-based data sets used to examine the genetic or biological basis of human races came from single linkage groups, such as mtDNA and the Y chromosome. But while trees constructed from these clonal markers are interesting in that they trace maternal and paternal lineages and are quite well resolved, as we pointed out in chapter 3 the bigger picture still needs to be kept in mind. Because while there will always be some general similarities between the maternal and paternal patterns of divergence within a species, they will also be at odds or incongruent with each other in many respects. This incongruence is difficult to reconcile; and indeed, there is no reason in principle to expect the patterns of divergence from the two linkage systems to agree. As we showed with Darwin's paternal phylogeny and maternal phylogeny in chapter 7, each has its own independent history.

The addition of whole genome–level information to the study of human genetics introduces additional massive complications, because now every linkage group in the human genome will have its own evolutionary history. How best to deal with this problem is controversial; but we will show in this chapter that any of the tree-based ways that are normally used will always result in unresolved trees, or in trees that make little sense with respect to any presupposed racial boundaries. To deliver the punch line right at the start, using reciprocal monophyly as a criterion for testing hypotheses about human races is a dead end.

The fact that different genes, chromosomes, and genomes have distinct evolutionary mechanisms and histories (are incongruent) is one of the reasons why tree-based analyses do not agree with methods that use a reticulate model. Two trees are incongruent when they do not agree in the phylogenetic inferences they yield, and there may be differing degrees of incongruence. Several tests have thus been proposed to determine the statistical significance of any incongruence observed between two trees and have showed just how pervasive incongruence is. One review of incongruence among alternative trees for diploid eukaryotic organisms revealed that between 22 percent and 50 percent of the single genes in a genome were incongruent with the taxonomically accepted topology for the species concerned. In one study of (very closely related) subspecies in the mouse species *Mus musculus,* only 33.3 percent of the genes analyzed agreed in their topology with the accepted genealogy for the subspecies. A further 33.3 percent agreed with a second topology, and the final 33.3 percent agreed with a third one. Effectively, this data set gave a random inference with respect to the accepted relationships.

The most obvious example of incongruence we have discussed so far involves the clonally inherited genes of mtDNA and the Y chromosome, which evolve without recombination. Because they do not recombine in each generation, the inferences made in the literature using these markers give well-resolved trees that are interpretable individually as histories of the markers involved. On the other hand, trees can be generated for each of the twenty thousand or so individual genes in the human genome, and some of them would also be resolved. But because of recombination and the lack of clonality of the individual genes, such trees will merely represent the history of each gene—recombination and all. If enough genomes are included as terminals in separate phylogenetic analysis of all the genes in the human genome (and the 1000 Genomes Project has sequenced more than two thousand of them), the probability that any two genes out of our twenty thousand or so genes would yield the same tree is extremely small, if not zero. This incongruence does not mean that gene trees, or the information from discrete gene regions, cannot be used to generate species trees; but at the infraspecies level the procedure would clearly be uninformative at best.

The problem of incongruence among gene trees is largely related to coalescent theory. Coalescent theory suggests that, even if there is a strong and irrefutable species tree, individual gene trees have a high likelihood of being incompatible. This observation suggests two very important things about gene trees and species trees. First, there is a basic theoretical and functional difference between a gene tree and a species tree. Second, the problem of unraveling the history of population divergence becomes one of inferring population or species trees from gene trees. One solution to the resulting problem of inference is just to ignore it and treat all the data as a single linkage group, concatenating the data to produce the phylogenetic tree. This strategy has a theoretical basis that hinges on hidden information in unlinked chunks of DNA. The idea is that two different chunks of DNA may give different trees, but there will be information in one that supports the other, and vice versa. So how do you know which is the better tree? The solution is to combine all the data and allow the overall signal in the data to emerge. It often turns out that neither tree is part of the best solution, so yet another tree is generated by this concatenated approach. An alternative solution uses coalescent analysis of multiple gene trees to come to a phylogenetic hypothesis, since it has been suggested that coalescent methods can extract the patterns supported by certain genes to give an overall phylogenetic hypothesis for a data set. But these approaches, too, have often been criticized as inaccurate indicators of the overall phylogenetic signals of organisms.

It has been our experience that, whatever approach you use for the human genome data, the trees generated are terribly unresolved. What is more, they never reflect preconceived "racial" or established geographical boundaries. Purely for the sake of argument, we will discuss the results using the concatenation approach. In a recent study one of us used phase 2 data from the 1000 Genomes Project for more than five hundred human males, employing mtDNA, Y-chromosome DNA, X-chromosome DNA, and chromosome 20 DNA. The reason males were used is that only males have Y chromosomes, so we could be sure that every individual in the study had been sequenced for all of the markers. Figure 11.1 (top) shows the phylogenetic trees obtained for the two clonal markers.

While there is some interesting structure to these trees, that structure basically mirrors what we have already shown for mtDNA and Y-chromosome DNA. The African lineages shown in black in the figure (see plate 9) are generally at the base of each tree, indicating that all human Y chromosomes and mtDNA diverged after humans left Africa. Close examination of the two trees indicates that there are no instances of reciprocal monophyly. One geographic group that does border on reciprocal monophyly is the Asian Y chromosomes; but this can be explained by a specific problem that phase 2 of the 1000 Genomes Project had with samples from people of Asian ancestry. This resided in the fact that the Asian lineages used are all very closely related to one another, whereas the African, European, and Amerindian lineages are more deeply divergent.

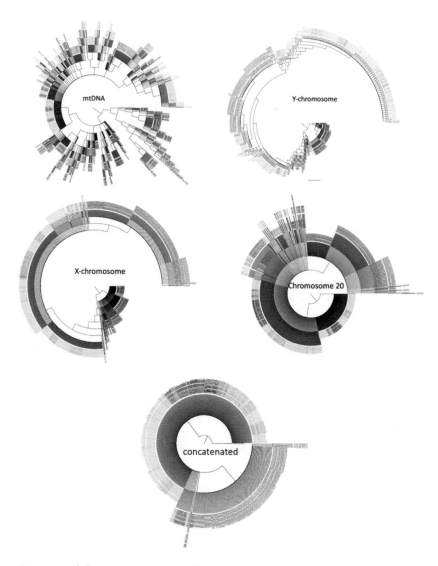

Figure 11.1 Phylogenetic trees generated from the single linkage group clonal DNA (*top*), recombining genetic blocks (*middle*), and concatenated data (*bottom*), all rooted with chimpanzee. The colors represent different geographic origins for the men in the analysis. Blue, Africa; red, Asia; green, Europe; orange, Amerindian. See plate 9.

If you think clonal genes are messy, then prepare yourself for an even bigger mess with recombining genetic blocks like chromosome 20 and the X chromosome (figure 11.1, middle). Reciprocal monophyly is nonexistent in these two trees (again, except for men of Asian descent) for recombining genomic elements. When all the data are concatenated, the tree in figure 11.1 (bottom) is obtained. Again, the only reciprocally monophyletic assemblage is the Asian one; but given the close affinity of the individuals from this geographic region, it is not surprising that they form a monophyletic group.

It should be clear that using the criterion of reciprocal monophyly produces no clear delineation of human males into groups and that the one exception is a sampling artifact in the data set from phase 2 of the 1000 Genomes Project. We would also predict that the picture will get even messier as more individuals are added to the analysis; it is a general characteristic of phylogenetic analyses that as the number of terminals added to a data set increases, the messier the structure of the tree will become. The addition of more and more variable genomes with different signals inevitably leads to decreasing resolution. Fair enough. But what if we simply reduce the amount of DNA in the analysis? Well, as we pointed out with ascertainment bias, trimming the data set will probably lead to better resolution in the tree; but while trimming can lead to cleaner groupings, it will also produce an unnatural answer to the initial problem.

By way of illustration, take the data set we presented in figure 8.3, and use only those SNPs that agree with the geographic grouping of the data. This is an extreme form of ascertainment bias that is directly related to the search for the ancestral informative markers (AIMs) we discussed in chapter 8 in the context of an example involving European populations. To recap, here is how AIMs are selected: sequences from people whom you think belong to different groups or "races" are lined up, and then scanned to find positions that are fixed and different in the various groups; that is, they diagnose those groups. Figure 8.3 shows this process and points to the determination of four AIMs, one diagnostic for each of the four groups. These AIMs are "pure" diagnostics, as there is complete fixation and differentiation of the AIM from one group to the next. In the real world, an AIM might not be completely fixed—although to be informative in any way it would have to be very close to fixation.

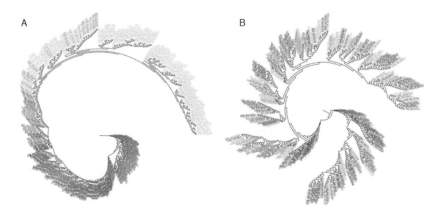

Figure 11.2 Phylogenetic analysis of AIMs (*A*) and anti-AIMs (*B*). See plate 10.

If we take AIMs as thus determined, and use only those SNPs to construct an evolutionary tree, we get an unsurprising result. When all the SNPs are concatenated together, a tree that looks as if it has monophyletic groups will be obtained (figure 11.2, left). But we have actually tricked ourselves into thinking this is a valid conclusion. For if we take the SNPs in the data set that are the least like AIMs and use them to make a phylogenetic tree, we will get the tree shown on the right side in figure 11.2—from which it is evident that the extreme ascertainment strategy used in AIMs is mislead-ing. To put it bluntly, it is based on circular reasoning. If you select markers that are fixed and different for a phylogenetic analysis, you will get the tree you want, because you have selected the data to fit your hypothesis. And as we will see in the next chapter, when other approaches such as clustering are used to analyze genome-level data for humans, similar problems arise.

Quite apart from the lack of resolution in these trees, and their inability to recover monophyletic groups that represent geographic races, there are strong theoretical reasons why building trees is not a valid approach for testing hypotheses about race. Early in the development of phylogenetic tree-building methods, Willi Hennig famously challenged the notion that trees could be used indiscriminately, using a figure that pinpointed the problems inherent in using tree-based approaches to describe biological systems (like species) that reticulate. ("Reticulation" is a word applied to branching diagrams in which there is connection among the lineages. In biological terms, it refers to interbreeding.) The level of biological organization that Hennig was concerned about in connection with reticulation is the level at which species form. In Hennig's view (and in line with the biological species concept, for that matter), before speciation, organisms of two populations will successfully interbreed with each other; afterward, they will not. Once there is a cessation of genetic contact, proxies for lack of interbreeding can be used to characterize the populations. But not before. It is difficult to deny that members of our species reticulate, which is of course what makes Hennig's ruminations so relevant here.

Systematics and population genetics both rely on distinct but equally entrenched ontologies and thus are completely different scientific "animals." In population genetics, the established laws of heredity are used to make evolutionary inferences from observations of populations. This is a deterministic approach that relies on the continuity of the evolutionary process, and its goal is to explain and detail the origin and subsequent maintenance of genetic diversity at the population level. Systematics, on the other hand, attempts to discover and describe the intricacies evident in the hierarchical pattern of life on this planet. This discipline is not a deterministic endeavor, but is rather a purely historical one. As Andrew Brower and colleagues suggest, "Systematics seeks to document hierarchical patterns among disjunct entities and needs to postulate little except that a tree-like hierarchy exists and is recoverable by studying attributes of individual organisms." This view has led to the suggestion that a "line of death" exists at this level of biological analysis. And indeed, the results of the phylogenetic analysis of SNP data in this chapter demonstrate some aspects of this line of death. One of the results of this unique boundary between population genetics and systematics, and of analyzing that slice in time when that boundary—speciation—occurs, is "Woodger's paradox."

John Woodger was a biologist who suggested in the mid-twentieth century that isolation as a criterion for speciation posed a paradox. Specifically, he claimed that, at the point during the speciation process at which divergence happens, a mother would have to be placed in species X, while her daughters would be in species Y. And the nature of species is such that a parent belonging to species X cannot give rise to offspring belonging to species Y. But there is, of course, a huge problem with Woodger's paradox. It makes two assumptions, each of which is fine in and of itself, but which in combination are lethal. The first assumption is that an individual cannot belong to two species; and the second is that the placement of an individual into a specific species is immutable with time. This led Woodger to view speciation—the emergence of new species that is an essential component in the diversification of life—as the drawing of a line between two successive generations and the result of some radical biological event. This leads to a view of speciation as a virtually instantaneous phenomenon, and one that is immutable with respect to time. Yet whatever it may consist of in any specific instance, speciation—a woefully poorly understood process or set of processes—clearly does not work like this. Time is an essential ingredient in speciation, and the emergence of reproductive incompatibility between two physically separated but nonetheless closely related populations can take quite a while—as evidenced, for example, by that putative hybridization between Neanderthals and modern humans. Accordingly, Woodger's basic assumptions about speciation were not optimal. His problem was that he attempted to understand speciation from one single perspective: that of population genetics. But when you hit that line of death, population genetics no longer works, and Woodger's conclusion is hugely problematic.

If we accept the argument about interbreeding populations, then we should eliminate trees as a valid way of representing reticulating individuals within a population. So although this does not necessarily mean that trees won't be useful at this level for other endeavors, we have to reject trees in the context of trying to understand human races. Unfortunately, however, even when researchers recognize this important limitation of trees in principle, they often continue to use them anyway. Why is this? The statistician J. C. Gower had a good answer in 1972: "the human mind distinguishes between different groups because there are correlated characters within the postulated groups". The underlying correlation of data that Gower mentions, and that so much beguiles us, brings us to clustering methodologies and to how researchers have used this approach to address biological race.

```
     m  1    2    3    4    5    6    7    8    9    0    1    2
Lion + ATG  AAT  TAT  ACA  AGT  TAT  ATC  TTA  GTT  TTT  CAG  CTC
Tiger + ATG  AAT  TAT  ACA  AGC  TAT  ATC  TTA  GCT  TTC  CAG  CTT
Bear  - ATG  AAT  TAC  ACA  AGT  TTT  ATT  TTC  GCT  TTT  CAG  CTT
Pango - ATG  AAT  TAC  ACA  AGT  TTT  ATT  TTC  GCT  TTT  CAG  CTT


     m  1    2    3    4    5    6    7    8    9    0    1    2
Lion + M    N    Y    T    S    F    I    F    A    F    Q    L
Tiger + M    N    Y    T    S    F    I    F    A    F    Q    L
Bear  - M    N    Y    T    S    Y    I    L    A    F    Q    L
Pango - M    N    Y    T    S    Y    I    L    V    F    Q    L
```

Plate 1

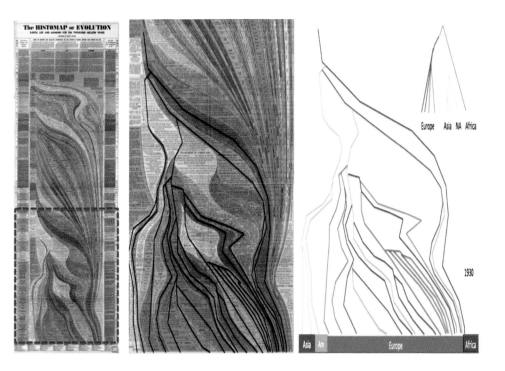

Plate 2. Sparks (1932) "Histomap of Evolution" is shown on the left. The red box at the bottom shows the part of the diagram dedicated to human evolution. The diagram in the middle shows the detail in the red box and represents the range of human evolution. We have traced the various groups of humans that Sparks follows with black lines. The diagram on the right is a tracing of the human lineages in the middle showing the groups to which the lineages belong.

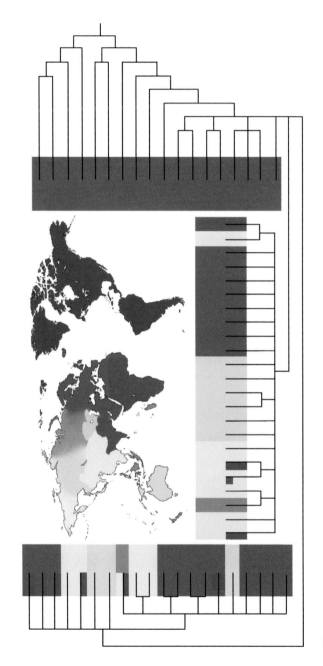

Plate 3

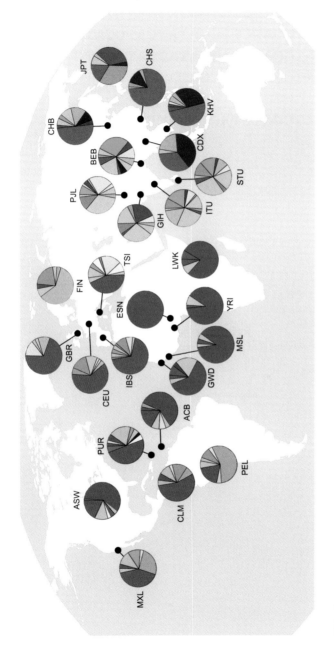

Plate 4

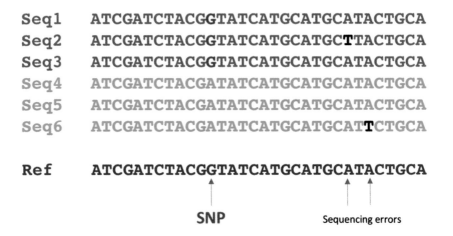

Plate 5

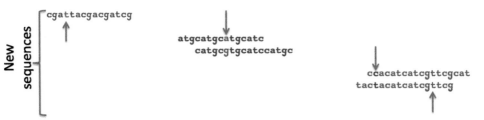

Plate 6

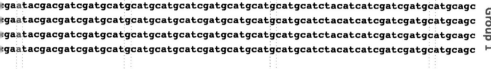

Plate 7

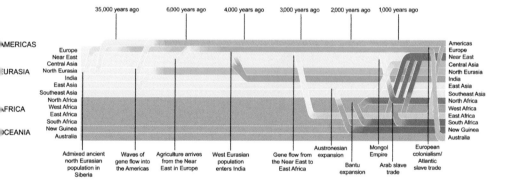

Plate 8

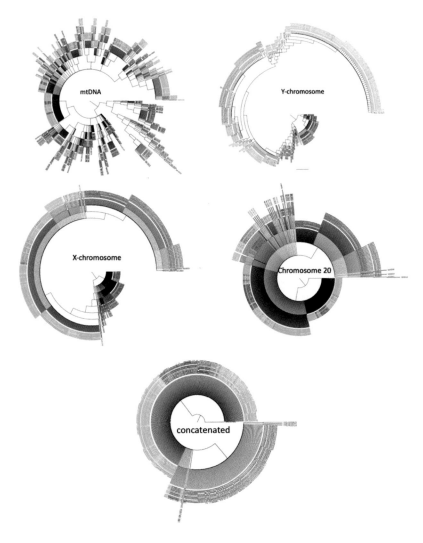

Plate 9

Plate 10

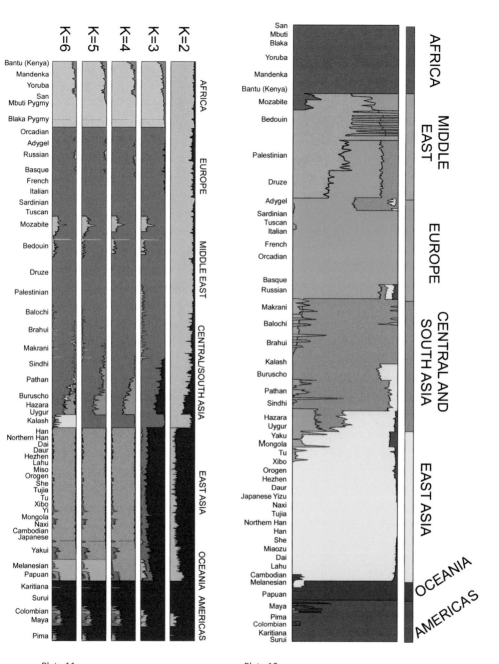

Plate 11   Plate 12

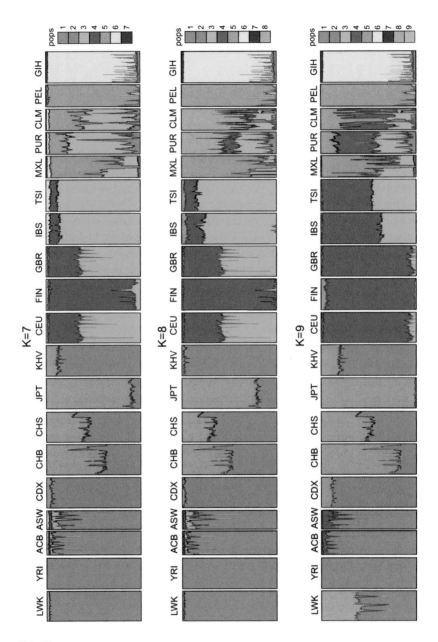

Plate 13

12

• • • •

Clustering Humans?

We are a curious species not only because of our unusual dichotomous way of viewing and understanding the world, but also because of how highly visual we are. If there is a visual way of presenting something, that is what most people will normally prefer. This goes for scientists, too, especially when large quantities of complex data are involved. Perhaps the most important question that scientists ask when confronted with such data is: "How are the data points distributed?" Visualizing such distributions then allows for a wide range of other questions to be addressed, such as "Is there a difference between one group of data points and another?," or "How similar is this group to that one?," or "How many groups of data points are there?"

The primary way to present data is to group them in a table. Tables in scientific papers go back a long way, and great examples come from the many scientific papers of Francis Galton, a late nineteenth-century cousin of Charles Darwin and a polymath in his own right. Along with the occasional regression line (a technique that he and the statistician Karl Pearson invented), the majority of papers Galton wrote about genetics and eugenics are most remarkable for their tables. Our favorite is the table in an oddly titled paper, "Statistical Inquiries into the Efficacy of Prayer". We reproduce this as figure 12.1.

Galton's tables were blurs of raw data and averages. But once he had introduced statistical ways of looking at such diverse subjects as prayer,

	Number.	Average.	Eminent men.[1]
Members of royal houses	97	64.04	
Clergy	945	69.49	66.42
Lawyers	294	68.14	66.51
Medical profession	244	67.31	67.07
English aristocracy	1,179	67.31	
Gentry.	1,632	70.22	
Trade and commerce	513	68.74	
Officers in the royal navy . . .	366	68.40	
English literature and science . .	395	67.55	65.22
Officers of the army.	569	67.07	
Fine arts	239	65.96	64.74

Figure 12.1 Galton's table from "Statistical Inquiries into the Efficacy of Prayer," published in 1872. Galton's test is simple. Since individuals in the 1800s were always praying for the long life of members of royal houses, he wanted to see whether these prayers did indeed confer long life on the royals. The first column represents the number of individuals Galton had information for in each category. The second column is the average life span of the category. The third column is the average age of eminent men in the indicated categories. Apparently royals live shorter lives (the only nonroyal category anywhere close are eminent men in the fine arts), hence the efficacy of prayer is somewhat in doubt!

stature, mental imagery, personality, and the "wisdom of crowds" (voting)— to name just a few—other researchers began looking for ways of presenting complex data that would be easier on the eyes. Some simply colored or shaded cells in data matrices to represent specific ranges of values. Others thought that rearranging the order of entries in the data matrices (much like doing a Sort in Excel) was a valid way to enhance the visual appreciation of complex data. Later in the development of data matrix visualization, researchers started to place branching diagrams on their matrices to represent how similar different classes of data were to others. All such approaches are forerunners of the "heat map" approach that has become an important tool in visualizing complex data from genomics studies. We must warn you, of course, that the branching diagrams in figures like these do not represent phylogenetic trees. Rather, they are simply a visual way of representing the similarities among the entries in the matrices.

What these diagrams are trying to do is cluster entities in the matrix based on the similarities of the measurements used to construct it. In the early part of the twentieth century, clustering became a popular approach among behaviorists, anthropologists, and social scientists. The first attempt to do this in psychology was made by Joseph Zubin in 1938. Zubin's interest was in what he called "like-mindedness." He wanted to be able to assess the similarity of people's personalities through their answers to a series of series of questions that placed his subjects into different personality clusters. Visually, he opted for tables to express his data and to demonstrate the validity of his conclusion.

The anthropologists H. E. Driver and A. L. Kroeber (1932) also described clustering methods, and they used several cultural anthropological subjects that were of great current interest at the time (Polynesians, Plains Indians, Northwest Coast Indians, and Northeast Peru Indians). Their clustering techniques were developed to address the question of whether there was an adequate signal in cultural anthropological data to allow relationships among cultures to be deduced. Driver and Kroeber described their methods as "so simple as to be intelligible with a knowledge of nothing more than arithmetic, and so humble as to have been generally over-looked by inquiries into statistical theory and little used by biological, psychological, and economic statisticians." They used similarity measures to come up with figure 12.2 for the Polynesian data set they examined, and they introduced an important difference from Zubin's study in visualizing their results: two-dimensional space. Along with the branching diagrams, this two-dimensional diagram has become the hallmark of clustering analysis.

In the 1950s Robert Sokal and Peter Sneath took this approach to its logical systematic extreme and suggested that clustering would be a valid way to describe the relationships of organisms. Their *Numerical Taxonomy* proposed an objective, operational, and rapid method for systematizing organisms in taxonomic research (Sokal and Sneath 1963). We presented this approach in chapter 3, but let's take a moment to look a little more closely here. First, as we noted, similarity and clustering have been shown to be crude and sometimes inapplicable when doing taxonomy and systematics. Think back to the goals of systematics, one of which is to obtain diagnostics. And it is easily shown that similarity and clustering do not lead to identifying characters that would be capable of diagnosing an individual to a species. Sure, one could say that group A is diagnosed as distinct from group B, with a similarity of 0.12. But the similarity is not a character

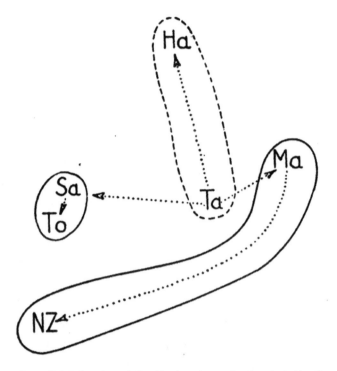

Figure 12.2 Polynesian relationships based on cultural context. The distances are those on the map. Cultural similarities are indicated by enclosures of different strength; probable courses of cultural relationship are indicated by arrows. Ha, Hawaii; Ma, Marquesas; NZ, New Zealand; Sa, Samoa; To, Tonga.

in the classical sense of the term. The similarity measure of 0.12 is not an inherent property of the individual; rather, it is an emergent property of comparing group A with group B. In contrast, if instead one diagnoses group A as distinct from group B because the diagnostic characters of the two are fixed and different, the character composing the diagnostic will not only emerge from comparing group A with group B, but it will also be inherent to individuals—which is what we need. Today, the "phenetic" methods described by Sokal and Sneath are rarely used for systematic and phylogenetic tree construction. Some systematists go so far (in our opinion rightly) as to reject the resulting branching diagrams as phylogenetic at all; and indeed we would rather see such figures characterized purely as "dendrograms," or visual summary diagrams that represent similarity alone. Still, visually, it is readily apparent why scientists like to use them. It is just plain easier on the eyes to look at a dendrogram than at a pairwise similarity matrix.

One of the more widely used approaches in this realm is principal components analysis (PCA). This approach, invented by Karl Pearson in 1901, has been subsequently rediscovered, renamed, and re-employed in fields as disparate as mechanical engineering, psychometrics, and linear algebra. PCA was invented to simplify the appearance of complex data, and it can be used to reduce the number of variables in a data set by discovering the variables that contribute the most to the structure in the data. The idea is that by re-ordinating data in space, one can reduce the variation in a data set by discarding the least informative parts of that variation. For instance, if there are two variables in a data set (as in the simple example below), then one of them (the one explaining the least amount of variation) can be dropped out of the analysis once the first variable is rescaled. The best way to do this is to determine the variables that explain the most variance in the data. Consider the following data on the height (ht) and weight (wt) of five individuals:

HT	WT
7	4
9	5
8	5
1	4
2	5

More than likely, a glance at this simple table won't tell you much. You could scan the data and come up with some interpretation, but we could much more effectively plot the data on a two-dimensional graph as in figure 12.3A. Since there are two variables (height and weight), PCA will reduce the representation of the data to one dimension, significantly simplifying the problem of visually representing the data. Determining the variance involves rotating the data in two dimensions (for this problem) or

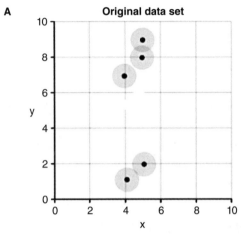

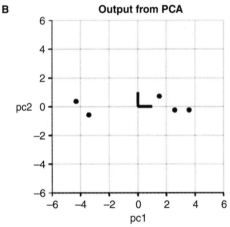

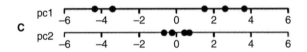

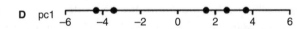

Figure 12.3 (A) Two-dimensional plot of the data in the two-column height/ weight table. (B) PCA plot of the same data. Note that the scale has changed (because variance is now plotted) and that the data appear to have been rotated 90 degrees. The degree of rotation is dependent on the variance of the measurements. (C) Plot of principal component 1 (PC1) and principal component 2 (PC2). Note that PC2 has much less variance than PC1 and hence is dropped, as the process's goal is to eliminate the visual effect of the variable with the least principal component variance. (D) Final plot of PC1 after the dimension reduction of PC2.

in *n* dimensions (for more complicated data matrices). For the same data set, this process results in the PCA plot in figure 12.3B.

Plots of the variance of PC1 and PC2 are shown in figure 12.3C. Note that PC2 has much less variance than PC1. In this example then we can "ignore" PC2 (also known as "dimension reduction"), so we are left with the plot in figure 12.3D, which is pretty easy to understand. Many find dimension reduction both desirable and defensible, because it results in a visually understandable and computationally tractable data set with minimal loss of information. Let's add a third set of data points to this example by adding the width of the individual (wd):

HT	WT	WD
7	4	4
9	5	4
8	5	4
1	4	3
2	5	3

Plotting these data in three dimensions leads to figure 12.4A. The PCA analysis of these three dimensions of data is shown pairwise in figure 12.4A. The PCA analysis is shown in the next three graphs (figure 12.4A–C). PC3 contains the least variance, as shown by figure 12.4B and C. The final graph demonstrates the dimension reduction (out with the width axis!) in which just PC1 and PC2 are left.

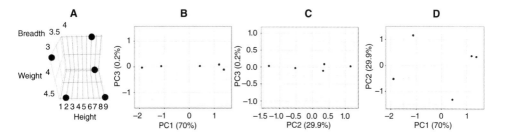

Figure 12.4 (*A*) Three-dimensional plot of the three-column data table shown in the text (height, weight, width); *x*-axis = height (ht); *y*-axis = weight (wt); *z*-axis = width (wd). (*B*) Graph of PC1 (ht) versus PC3 (wd). (*C*) Graph of PC2 (wt) versus PC3 (wd). (*D*) Final dimension reduction where PC1 (ht) and PC2 (wt) are graphed.

When the variance explained by the PCA for each of the variables shows large gaps—as it does in the examples shown here—researchers generally agree that dropping the least informative variable (dimension reduction) is a reasonable step. Olivier François and colleagues recently reviewed the origin of the PCA approach in human population genetics. Their justification for using the approach is mostly operational, and their description of PCA is interesting and might help clarify how the approach is used in human population genetics:

> One way to explain PCA is as an algorithm that iteratively searches for orthogonal axes, described as linear combinations of multivariate observations, along which projected objects show the highest variance, and then returns the positions of objects along those axes (the principal components [PCs]). For many data sets, the relative position of these objects (e.g., individuals) along the first few PCs provides a reasonable approximation of the covariance pattern among individuals in the larger data set. As a result, the first few PC values are often used to explore the structure of variation in the sample. (François et al. 2010, 1257)

One of the side effects of PCA is that an n-dimensional data set can be reduced to a one- or two- or three-dimensional picture through the process of dimension reduction. Plots of the data (usually with two or three dimensions at the end of the day) often show clustering of data points. Look again at figure 12.3D, and note that visually there is a gap between the two points on the left and the three points on the right in this one-dimensional graph. Now, imagine a data set with thousands of data points. While the tables so far presented in this chapter might be a little rough on the eyes, a matrix with thousands of entries will, we guarantee you, be many times worse. When the PCA method of dimension reduction is applied, most of the dimensions in the data set can be brushed aside, and a visually understandable presentation of the data can be made. One of the questions we should address is the number of dimensions we need to make sure the dimension reduction doesn't render the analysis useless, because the PCA will no longer provide a good representation of the data. For our two examples (figures 12.3 and 12.4), this problem is trivial, because the one-dimensional plot in figure 12.3D represents more than 90 percent of the variance in the data, and the two-dimensional plot in figure 12.3D represents about 99 percent of the variance in the data set.

PCA has been used in studies of human variation since the early 1960s, and the omnipresent Luca Cavalli-Sforza and his colleagues (e.g, Cavalli-Sforza, Menozzi, and Piazza 1993) used the approach in their seminal and broadly influential work in the 1990s on the history of human genes. PCA in the Cavalli-Sforza studies resulted in synthetic maps (PC maps) that are essentially PCAs superimposed on the geographic regions where the samples were collected. In many of the loci examined, and in many of the PC maps generated using allozyme data, gradients of allele frequencies are clearly observed.

When genomics methods became available, PCA seemed like the perfect technique to reduce the noise in the amazingly complex data of the genome. PCA is widely used in human genomics, and our review of the literature revealed at least thirty-three large-scale genomic studies that use the technique published in the last five years. Many of these studies used the paleogenomes of Neolithic individuals or focused on the clustering of individuals within specific large geographic areas like Africa, Europe, the Americas, and Asia. Several of the studies used PCA to examine the structure of localized or genetically related populations (such as Jewish populations and regional populations from Sweden, Columbia, Ireland, Japan, and India, to name only a few). Several of the studies addressed worldwide patterns, and these are the ones we will focus on here.

Table 12.1 summarizes nine such studies. It shows that there are many ways of using PCA to discover clusters and that there is also some degree of variation in the number of clusters that will be discovered using PCA. Depending on how the PCA is constrained, this approach will yield anywhere from three to nine global clusters. The analyses are dependent both on sample size (N), and on the number of populations used in the analysis. More important is the method by which the SNPs used in the study are ascertained. By using only AIM-based SNPs, one can increase the amount

TABLE 12.1
Summary of Genome Studies That Use Worldwide Sampling and PCA

Reference	N	Pops.	No. of SNPs	%PC1/PC2	Clusters	Note
Li et al. (1)	**938**	**51**	*	**86**	**4 or 5**	**Fst was used for 51 × 51 pairwise population comparisons.**
Paschou et al. (2)	**255**	**11**	10,805	45	4	**30 PCA-correlated SNPs were used to characterize relatedness.**
Auton et al. (3)	3,845	11	443,434	7	7 or 8	The analysis resulted in overlapping clusters.
Nassir et al. (4)	**1,620**	**20**	93 AIMs	80	6 or 7	**Authors used 93 SNP AIMs to generate clusters.**
Sudmant et al. (5)	2,504	26	68,818	8	3 or 4	Genomes from Africa and Asia were taken out, and the result is that other populations overlap.
Biswas et al. (6)	1,043	52	46,000	<10	6 or 7	Authors used SNPs correlated to PC to reduce data set.
Lek et al. (7)	60,706	5	5,400 c/e	NA	5	5,400 common exome SNPs were used to generate clusters.
Martin et al. (8)	2,504	26	Filtered	14	3	The genomes in the data set showed extreme admixture.
Cherni et al. (9)	**2,984**	**57**	399	87	**5–9**	**Authors used 399 handpicked SNPs to generate clusters.**

*This study is unique in that it computes the Fst (a measure of differentiation) based on 650,000 SNPs to use in the PCA, which reduces the data set considerably.

Studies cited: (1) Li, Jun Z., Devin M. Absher, Hua Tang, Audrey M. Southwick, Amanda M. Casto, Sohini Ramachandran, Howard M. Cann et al. "Worldwide Human Relationships Inferred from Genome-Wide Patterns of Variation." *Science* 319, no. 5866 (2008): 1100–1104;

(2) Paschou, Peristera, Elad Ziv, Esteban G. Burchard, Shweta Choudhry, William Rodriguez-Cintron, Michael W. Mahoney, and Petros Drineas. "PCA-correlated SNPs for Structure Identification in Worldwide Human Populations." *PLoS genetics* 3, no. 9 (2007): e160;

(3) Auton, A.,, K. Bryc, A. R. Boyko, K. E. Lohmueller, J. Novembre, A. Reynolds, A. Indap et al. "Global Distribution of Genomic Diversity Underscores Rich Complex History of Continental Human Populations." *Genome research* 19, no. 5 (2009): 795–803;

(4) Nassir, R., R. Kosoy, C. Tian, P. A. White, L. M. Butler, G. Silva, R. Kittles et al. "An Ancestry Informative Marker Set for Determining Continental Origin: Validation and Extension Using Human Genome Diversity Panels." *BMC genetics* 10, no. 1 (2009): 39;

(5) Sudmant, P. H., T. Rausch, E. J. Gardner, R. E. Handsaker, A. Abyzov, J. Huddleston, Y. Zhang et al. "An Integrated Map of Structural Variation in 2,504 Human Genomes." *Nature* 526, no. 7571 (2015): 75–81;

(6) Biswas, S., L. B. Scheinfeldt, and J. M. Akey. "Genome-Wide Insights into the Patterns and Determinants of Fine-Scale Population Structure in Humans." *American Journal of Human Genetics* 84, no. 5 (2009): 641–650;

(7) Lek, M., K. J. Karczewski, E. V. Minikel, K. E. Samocha, E. Banks, T. Fennell, A. H. O'Donnell-Luria et al. "Analysis of Protein-Coding Genetic Variation in 60,706 Humans." *Nature* 536, no. 7616 (2016): 285–291;

(8) Martin, R. et al., 1000 Genomes Project Consortium. "A Global Reference for Human Genetic Variation." *Nature* 526, no. 7571 (2015): 68–74;

(9) Cherni, L., A. J. Pakstis, S. Boussetta, S. Elkamel, S. Frigi, H. K. El-Khil, A. Barton et al. "Genetic Variation in Tunisia in the Context of Human Diversity worldwide." *American Journal of Physical Anthropology* 161, no. 1 (2016): 62–71.

of variance explained by the first two PCs. For instance, the studies in table 12.1 that use reduced or AIM-ascertained SNPs (bold entries) can explain up to 87 percent of the variance in the data. This makes sense, because the removal of SNPs that are not AIMs removes most of the noise in the data set and allows the remaining SNPs to explain most of the variance. On the other hand, when all the SNPs are analyzed together and allowed to influence the variance in the data, less than 10 percent of the variance is explained by PC1 and PC2 (the components that explain the most and second-most variance).

Clustering is evident in most of these studies, but it is difficult to come to an overall interpretation of the data with respect to the reality of geographic or racial clustering. This is because of the subjectivity with which some of the studies ascertain their data (bold entries in table 12.1). When there is a higher level of objectivity (entries in table 12.1 that are not bold), then far less of the variance in the data set can be explained. While not quantifiable, there is a distinct decrease in discrete clustering when higher objectivity is used to ascertain SNPs in such studies.

Some improvements on PCA have been made recently to shore up the statistical aspects of the approach. For instance, David Reich and colleagues have developed a process called EIGENSTRAT to determine the statistical significance of the PCs generated in a PCA (Price et al. 2006). The name "EIGENSTRAT" comes from the use of mathematical eigenvectors to stratify (group) individuals in genome-wide association studies. For our two simple examples in this chapter, using only two or three variables, this is a straightforward process of asking whether the degree of variance explained by the PCs is meaningful. For both examples, PC1 is highly significant statistically, meaning that the dimension reduction picture it gives us is probably an accurate one. However, in most of the PCA examples we have cited, statistical significance was not determined using this approach. One of the important applications of EIGENSTRAT is that the PCs most statistically significant for a data set can then be used as markers for the structure of the data set concerned.

These observations beg the question of what the PCA analyses are telling us about differentiation within our species. Here we can turn to

Anil K. Jain, a researcher using PCA and clustering-based analysis in non-genetic contexts. His perspective is illuminating. Of clustering, he says:

> Organizing data into sensible groupings arises naturally in many scientific fields. It is, therefore, not surprising to see the continued popularity of data clustering. It is important to remember that cluster analysis is an exploratory tool; the output of clustering algorithms only suggests hypotheses. (Jain 2010, 651)

We submit that this is the best way to look at the results of any clustering analysis. While clustering techniques may be very useful in visualizing data and helping to formulate hypotheses, we should not expect them to give us any final answers or to usefully test any hypotheses. We suggest, therefore, that citing the results of PCA clustering as "evidence" for fully differentiated human populations is not a procedure that should be employed by any scientist who wishes to be regarded as credible. Many researchers who use PCA and other clustering approaches are well aware of this, of course. But the visual appeal of clustering is in many ways the siren song of racial research, so caution is always in order.

13

• • • •

STRUCTUREing Humans?

Dimension reduction methods using principal components analysis (PCA) lead to visual ways of clustering things; but clustering methods have also been applied to understanding the genomics of human population structure. Jonathan Pritchard, Matthew Stephens, and Peter Donnell point out that there are two major approaches to clustering: distance-based and model-based methods. Distance-based methods first convert the raw DNA sequence data into distances (or similarities). Similarity is simply the flip side of distance, so the approaches to computing similarities are intimately connected to those used to compute distances. It doesn't matter much which is used, but it is very important to be consistent and to know which one is being used when using coordinates or any distance/similarity approach.

One complication with using distances lies in the ways in which the raw data are manipulated into distances (or similarities). There are currently tens of ways in which such transformations can be made. And because not all are directly related or proportional to one another, significantly different results can be obtained depending on which distance transformation (similarity measure) is used. In all cases, though, once distances have been computed for all possible pairwise comparisons, some visual way of presenting them is selected. We have already discussed one way of displaying distances by using dendrograms, and we have pointed out the shortcomings of these tree-like figures as indicators of phylogeny. But dendrograms are a nonetheless a pretty good way to reduce dimensions and visualize distances, because a dendrogram produced from a distance matrix is easily inspected by eye for any suggestion of clustering.

A second approach to visualizing distances is principal coordinates analysis (PCoA), which proceeds in a manner broadly similar to PCA. Both approaches produce a two- or three-dimensional graph, with the items of interest appearing as points in the graph. The difference is that instead of using the raw data as in PCA, distances between the items of interest are computed. All pairwise distances among the items of interest in the analysis are used, and when genomic data is being used, those points of interest are genomes or individual people. Those distances are then projected onto a multidimensional space, and the PCoA analysis begins by placing the first item of interest at the origin of the space. The second item of interest is then placed on the second axis at its proper distance from the first point, the third item at its appropriate place along the third dimensional axis, and so on for as many dimensions as there are items to be analyzed. As in PCA, a dimension-reduction step is then necessary, and this is accomplished in the same way in both techniques. PCoA, also known as multidimensional scaling or nonmetric multidimensional scaling, is considered by many to be a preferable approach to clustering compared with PCA, and it is used extensively in ecological studies.

Distance-based approaches have the advantage of being visually appealing, probably because they oversimplify the complexity of the data. Since this can be done in various ways, the overall visual product is, as we have noted, highly contingent on the distance transformation and the way in which the graphical representation is implemented. Indeed, differences among methods may be so great that, as Pritchard and colleagues also point out, it is difficult to assess statistically how confident we can be concerning the clusters. In fact, these authors suggest that the distance-based approaches are best for simply exploring the data but not good at all for making statistical inferences. This is where model-based clustering approaches come in.

Model-based approaches to clustering start with the basic assumption that the clusters themselves are made up of observations that can be considered randomly drawn from a statistically definable distribution. The

parameters of the actual clusters can then be inferred by implementing statistical methods that use descriptive models of those distributions. These
statistical methods are taken from the likelihood and Bayesian approaches
we discussed in chapter 3. Even though these model-based approaches
take the statistical properties of the data into consideration, there remains
the issue of visualizing complex data sets such as those presented by the
human genome. And problems may arise when the analytical approach
says one thing, while the visualization may be interpreted to say another.
This is the case for a rather colorful model-based analytical approach called
STRUCTURE.

STRUCTURE is a Bayesian approach developed by Pritchard and colleagues to handle the analysis of the burgeoning genome-level information becoming available for populations of organisms. Remember that
Bayesian approaches do not directly search for answers or solutions, but
rather seek to characterize and generate distributions from which probabilities can be calculated. STRUCTURE, then, seeks to generate a distribution of solutions for a likelihood model of population structure. Once
that distribution is generated, statistical tests can be performed on various
parameters of the model. One important set of parameters that can help
characterize clusters concerns the frequencies of alleles or SNPs in particular populations. Another very important parameter is the number of populations in the model, its "population structure," represented by the letter K.
Once the statistics of these parameters are established, it is a simple matter
to obtain a statistical statement about K, the number of clusters in the
data set. Once the number and identity of the populations is known, each
individual in the data set can be characterized by population. Other critical parameters of the model include allowing for admixture (interbreeding
of populations) or not; and contrasting the results of these two opposite
assumptions can give us a picture of the admixture of individuals in the
populations. Such "assignment" can lead to simple inferences about the
populations to which individuals belong.

For example, one individual might be assigned to a specific population
X. Another individual might be assigned with a 100 percent probability
to another population, Y. Since this is a statistical approach, a third individual might have a 0.3 probability of belonging to population X and

a 0.7 probability of belonging to population Y. Graphical representations of these three individuals are given in figure 13.1. Since this step can be applied to each individual in the populations studied, and for all K clusters, the visual presentation of such analyses often uses shading and color extensively (see plates 11 through 13).

One problem with using structure is that it is clunky, and making inferences using this approach takes a lot of computer time. Hence David Alexander, John Novembre, and Kenneth Lange developed a fast algorithm that uses the likelihood model of STRUCTURE to make rapid estimates of K and the admixture of individuals in a data set. They call their approach ADMIXTURE, and its output looks pretty much identical to that of STRUCTURE (figure 13.1). And while speed of computation should never take primacy over accuracy in the choice of analytical tool, it does appear that ADMIXTURE does a pretty good job of rapidly analyzing high-density genome-level data.

Finally, some researchers have attempted to combine the distance-based approaches and the model-based approaches to make inferences about population structure. Specifically, Daniel Lawson and colleagues have created an algorithm called fineSTRUCTURE that attempts to focus on the information about population structure that is contained in how SNP variation is distributed along chromosomes. After all, the SNPs do come from specific regions of the chromosome, and including this information gives

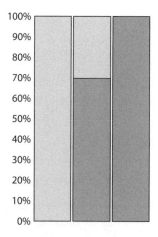

Figure 13.1 Bar plots for three individuals in a STRUCTURE analysis who have been assigned to two different populations (light shade is population X; dark shade is population Y).

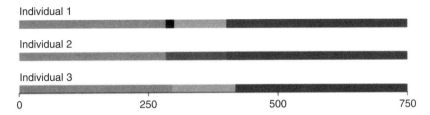

Figure 13.2 Chromosome painting showing three individuals whose chromosomes have been painted. The numbers at the bottom of the diagram indicate the chromosomal location of the painting. There are five different ancestral chromosomal "chunks" represented by different shades.

an analysis an added dimension. Specific patterns on chromosomes can be used as haplotypes that are viewed as coming from specified ancestors. The algorithm can then "paint" chromosomes with these ancestral haplotypes. Indeed, the process is called "chromosome painting," and like STRUCTURE it gives a rather colorful summary of the data, as shown (in grayscale) in figure 13.2.

The different chromosomal patterns can then be statistically dissected, and individuals can be compared to give what Lawson and colleagues call a "coancestry matrix." The coancestry matrix can be used to accomplish PCA-like and STRUCTURE-like inferences. One of the shortcomings of STRUCTURE and ADMIXTURE analyses is that the estimation of K is most accurate when K < 10. fineSTRUCTURE, in contrast, can be used to dissect hundreds of clusters in high-density genomic data.

STRUCTURE has been cited in nearly twenty thousand publications since its first appearance in 2000. The ADMIXTURE algorithm has existed since 2009, and as of June 2017, it had been cited more than one thousand times. Many of these studies examined human population structure, and some have become notorious, because they were subsequently enlisted by people with political agendas as proof of genetic reality of race. So it is worthwhile to look at some of the more prominent among them here.

While the STRUCTURE algorithm was first used to examined a bird system (the Taita thrush), in their original paper in 2000, Pritchard and colleagues also examined human population structure. At that point in time, very limited whole-genome sequencing had been done on human populations, so the researchers used a small data set of thirty SNPs generated using the old-school method of restriction fragment length polymorphism (RFLP) that we described in chapter 7. The individuals in the study included 72 people of African ancestry and 90 of European ancestry. Pritchard and colleagues looked at the probabilities of K = 1 and K = 2, and determined that K = 2 was very significantly more probable than K = 1, indicating the existence of two distinct populations. But does the story finish there? Using the same approach, Pritchard and colleagues then showed that K = 3, K = 4, and K = 5 might all have higher probabilities than K = 2. And indeed, the second result appears to be a recurring theme in those studies where K is the focus of attention. That is, upon further examination, the Ks of various values are all shown to have probabilities close to the one with the highest probability. Another area of concern about this first study of human populations using STRUCTURE involves the rather paltry sample size and ascertainment of the genetic data used to do the analysis. All things considered, while some very specific interesting questions were addressed with this early data set, applying its results to understanding human population structure was something of a stretch.

Two studies cited by the journalist Nicholas Wade in his 2015 book, *A Troublesome Inheritance: Genes, Race and Human History*, used the structure approach on more refined genomic information. The first of these was published in 2002 by Noah Rosenberg and several colleagues. They used 377 genetic markers, chosen to maximize the potential for discovering variability, on 1,056 individuals from 52 discrete populations. The results of their STRUCTURE analysis, shown in figure 13.3, led the authors to suggest that there are six human populations based on the statistics of the various K values examined.

Interestingly, five of the six populations coincide with geographic groupings of people, and this is what Wade focused on in his explanation of genetics and race. To shore up his thinking, Wade cited another paper by Rosenberg and colleagues (Ramachandran et al. 2005), in which they slightly reduced the sample size (down to 1,048 from 1,056) but tripled the number of genetic markers to 993. Not surprisingly (at least to us), the results were pretty much the same as with the 2002 data set, with K = 6 once again the statistically preferred value.

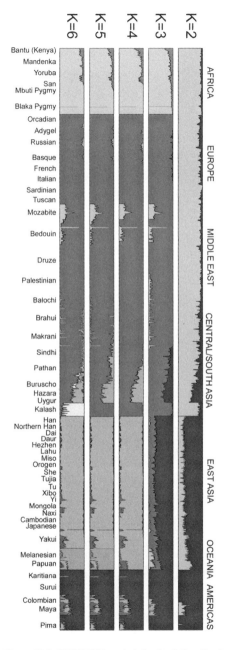

Figure 13.3 STRUCTURE analysis for K = 2 thru K = 6 from the Rosenberg et al. (2002) data set. K = 6 is statistically significantly better than all other Ks. Redrawn from Rosenberg et al. (2002). See plate 11.

These analyses led Wade to claim that five definable races of humans exist, a conclusion we will examine in detail in the next chapter. For now, we will just point out that the same sample size problem existed with the study published in 2005 by Rosenberg and colleagues as had marred their original study three years earlier. Merely piling on more loci that had been ascertained in the same way as the first 377 markers did not strengthen the inference at all. In a third study that has been cited as proof of six or seven "races" of humans, Junzhi Li and colleagues upped the ante of loci by several orders of magnitude in 2008. They examined over 600,000 SNPs (figure 13.4); and again, not surprisingly, they obtained K = 7 as the statistically preferred indicator of population structure.

Since 2008 many papers have been published using STRUCTURE, ADMIXTURE, and fineSTRUCTURE to address the burgeoning human whole-genome data of the last few years. It would be painfully redundant for us to go through all of them. Suffice it to say that the last five years have witnessed two major kinds of studies using these computer programs. The first are fine-grained studies that look at localized populations such as Ashkenazi Jews, British people, Tibetan highlanders, and sub-Saharan Africans. The second category consists of studies that take a global look at all human population structure. Both kinds of study have had to deal with expanding sample sizes, rapidly enlarging upward from the 1,000 to 2,000 individuals in the 1000 Genomes Project and the HGDP. As you might imagine, computation times for these larger data sets has been a problem, and several tools have been developed to handle this issue. Prem Gopalan and coworkers call this "tera-sampling," and have developed a tool called teraSTRUCTURE to handle the tera-data. In figure 13.5 we present results from their analysis of 1,718 people, at 1,854,622 SNPs. Note that, as the number of people included in the STRUCTURE analysis goes from 1,000 in 2008 to 1,700 in 2015, more and more of the solid colored blocks from the 2008 analysis start to bleed into colors from other blocks.

You can certainly look at the diagrams in figure 13.5 and see divisions; but to view them as indicators of "racial" separation is a bit of a leap. In addition, various studies over the past five years have shown that, as more people are added, K changes. Accordingly, Sarah Tishkoff and colleagues have shown that K leaped from 7 to 13 through the addition of a handful

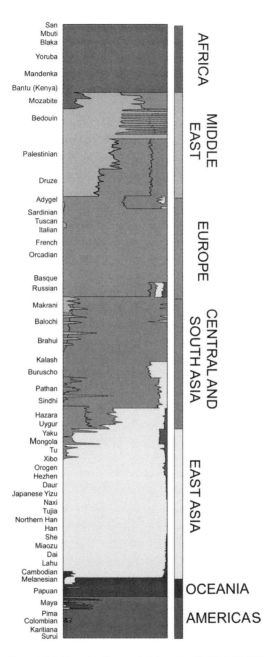

Figure 13.4 Li and colleagues' data set with 600,000 SNP loci and almost 1,000 human genomes. Only the statistically significant K = 7 is shown in this figure. Redrawn from Li et al. (2008). See plate 12.

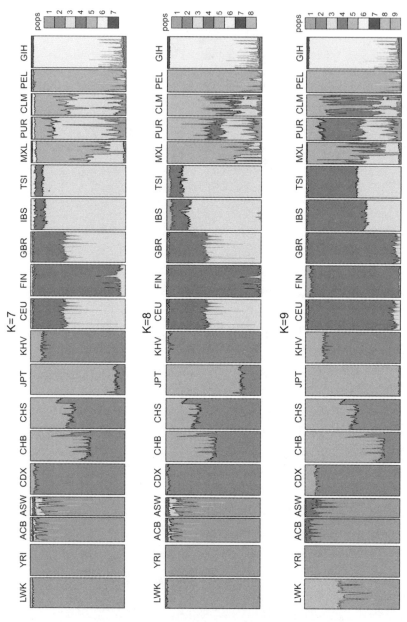

Figure 13.5 teraSTRUCTURE analysis of the 1000 Genomes Project data set (also known as TGP). The data set has 1,718 genomes at 1,854,622 SNPs. The top diagram is for K = 7, the middle is for K = 8 and the bottom is for K = 9. Redrawn from Lawson et al. (2012). See plate 13.

of African genomes (Bryc et al. 2010). Adding even more genomes will probably thoroughly bleach out many of the apparent divisions in human population structure.

Another problem with analyses like these, and a more serious one, is raised by John Novembre and Benjamin Peter's statement about the visual appeal of these approaches. In their review of human population structure, these scientists cite an important caveat in the use of these tools in studying human population structure:

A final precaution, and one of broader societal relevance, is that a viewer can become misled about the depth of population structure when casually inspecting visualizations using methods such as PCA, ADMIXTURE, or fineSTRUCTURE. For example, untrained eyes may overinterpret population clusters in a PCA plot as a signature of deep, absolute levels of differentiation with relevance for phenotypic differentiation. (Novembre and Peter, 2016, p. 102).

This sage warning takes us back to J. C. Gower's admonition that "the human mind distinguishes between different groups because there are correlated characters within the postulated groups" (Gower 1972, 11). It is to those correlated characters that we now turn.

The underlying correlation of data that Gower mentions evokes what A. W. F. Edwards called "Lewontin's fallacy." Recall that in the 1970s, based on genetic information, Richard Lewontin suggested that there is no hierarchy with respect to human populations. Indeed, he calculated that 85 percent of the variation in human genomes can be attributed to within-population variance, and only 15 percent to between-population variance. This between- and within-population variance pattern led Lewontin to state emphatically that between-population variance does not directly reflect separation of populations, or the strength of such separation.

Lewontin's findings lead to the following question, as posed by Lynne Jorde and colleagues: "How often is a pair of individuals from one population genetically more dissimilar than two individuals chosen from two different populations?" (Witherspoon et al. 2007, 357). The answer, it turns out, depends on the number of loci (or SNPs) used. It can be as much as

30 percent for low numbers of loci and limited population sampling, but with large numbers of loci the percentage drops to zero. However, Jorde and colleagues also point out that as more and more people are added to the base populations, the percentage will not be zero. The number also depends on how the loci are ascertained and, specifically, on whether common or rare allele loci are used. Rare allele loci will result in high frequencies of between-population similarities compared with within-population similarities.

Lewontin's observation of greater within-group variation is thus sustained. But what about Edwards's claim that Lewontin's conclusion ignores the correlation structure of data, and in so doing misses the fact that populations can be distinguished and delineated from one another using this underlying correlation structure? Well, modern phylogenetic systematics theory would not agree that a fallacy exists here. What Edwards is asking us to do is to ignore, or to "throw out," some of the information in the overall data set. In fact, a great deal of it. Modern phylogenetic systematics would urge us to search for a signal not only in the information that has underlying correlation structure but in the rest of the data set as well. Using only the information with underlying correlation structure is more akin to "clique analysis" or "compatibility analysis" in systematics. Clique analysis uses correlated characters to strengthen phylogenetic hypotheses, but it has been shown to be an unsound approach when applied to basic phylogenetic reconstruction. We can, moreover, say the same thing for the visual STRUCTURE and PCA approaches, in which the visual discreteness of divisions becomes reduced as more and more data are added.

If the ontological constraints of systematic analysis imply that the underlying correlation structure can't be used to extract systematic hierarchy or discreteness from human genomic data sets, what *can* it do for us? Well, context is everything here. Certainly, the underlying correlation structure is telling us something about ancestry. But whatever that may be, it is completely unlinked from our efforts to delineate the existence of populations within our species or to establish hierarchies of individuals. In the age of genomics, determining ancestry has become the identification of parts of the genome in two individuals that may be traced back to a common point in their histories. And while ancestry is a legitimate focus of studies using genome information about *Homo sapiens* on a large scale, such exploration of ancestry does not need to use phylogenetic trees, or clustering approaches. In fact, constructing trees and inferring clusters is downright illogical when the goal is to understand the ancestry of individual *H. sapiens*. A swivel toward ancestry, and away from hierarchy and clustering, would place the endeavor of human genome analysis on sounder philosophical grounds by eliminating the shaky assumption that predetermined population trees exist.

14

• • • •

Mr. Murray Loses His Bet

Nicholas Wade's recent book on the biology of human races, *A Troublesome Inheritance*, has by now been reviewed in many venues. We focus on the book and its conclusions here because it is the most recent attempt to racialize genomics and serves as an excellent example of the problems with such endeavors. The book has a simple structure. The first part argues that scientific orthodoxy can be stifling, and that to break away from it brave purveyors of truth must speak out. The second section argues that there is indeed genetic evidence for the biological basis of race. And the third suggests that, because there are races, we can now pinpoint a reason why different peoples purportedly behave differently. In his *Wall Street Journal* review of Wade's curious volume, the libertarian political scientist Charles Murray suggests that the book's last part would be the target of most criticism, reasoning that:

> The orthodoxy's clerisy will take that route, ransacking these chapters [the final five chapters] for material to accuse Mr. Wade of racism, pseudoscience, reliance on tainted sources, incompetence and evil intent. *You can bet on it* [italics added]. (Murray 2014)

Our intent here is to examine the science and premises in the first two parts (five chapters) of Wade's book. We can safely ignore its third part, because it can only be taken seriously if the premises in those first five chapters have scientific validity—and it is clear to us they do not.

Our reading of the first half of *A Troublesome Inheritance* indicates that Wade has made at least seven mistakes that are routinely committed when genomics and genetic information are used both to examine the biological basis for human races and when they are used as a justification for reifying race as a biological reality. We start with a foundational problem that all scientists face:

1. Misunderstanding the Nature of Hypothesis Testing

This first aspect of the "biology of race" controversy gets at the very core of what science is, and indeed of what the problems really are in understanding human variation. It is commonly accepted that the hypothetico-deductive (hypothesis-testing) approach provides the most sound and productive way to conduct science. In contrast, inductive approaches are to be avoided, because induction can only confirm what one already knows. This latter position might at first sound extreme; for if you have an approach that confirms a scientific phenomenon, why not use it? The answer is simple: science advances through the rejection of false hypotheses, while inductive processes will always give you a positive answer. Hence, with respect to "racial" variation in human populations, the proper approach is to pose hypotheses and subsequently test them.

As we saw in the last chapter, STRUCTURE is one of the methodologies most commonly applied in the analysis of human population genetic information. And unfortunately, it takes an entirely inductive approach, throwing data at an algorithm and asking: "How many units do I have?" Wade cites this method approvingly as the ultimate proof that there are five races of humankind. But while the algorithm itself is an important technical advance, how the results of such analyses bear on definitions of "race" is an entirely separate question because of the inquiry's inherently inductive nature. As we saw, inductive approaches can do a great job of summarizing and displaying information given prior knowledge of a system, and can thus encourage the formulation of new hypotheses and the refinement of existing ones. But what they cannot do is test those hypotheses.

To make scientific statements about race, then, we need to have hypotheses in hand, arrived at inductively or otherwise. So what useful hypotheses can we offer up with respect to human genetics and the existence of human races? The most obvious one is:

H_0 = There are n "races" of a type of organism (A) that correspond to the n geographical divisions (often taken to be Africa, Asia, and Europe) that we see on the planet today.

But simply posing our hypothesis in these terms brings us to the second problem with using biology to "prove" race.

2. Subjectivity in Defining Race
(or a Misunderstanding of What a Species Is)

How can we test a hypothesis of the kind we have just presented? First of all, we need a definition of race that is both objective and operationally testable. Without such a definition, we cannot proceed to test the hypothesis. We cannot ask an algorithm to give us an idea of the number of races, because that would be in itself inductive. We do have a good idea of what species themselves are; but the definitions of the subordinate units of race and subspecies are substantially less than objective. In fact, we defy any scientist, journalist, philosopher, or layperson to come up with a meaningful definition of race in this context and in such a way that it might be used to test our hypothesis H_0. And if this can't be done, H_0 becomes useless scientifically. However, you might change the hypothesis to:

H_1 = There are n species of a type of organism (A, B, and C) corresponding to the geographical divisions (for the sake of argument, Africa, Asia, and Europe) that we see on the planet today.

In this case we do have a testable hypothesis, because we do have an operational definition of species. You might object that this is just semantics. But in fact, objective definitions are hugely important in hypothesis testing. Without objective criteria to test our hypotheses, we simply cannot reject those hypotheses.

But then you might say, "I will objectively define a race as being a group differentiated from other closely related entities." This is slightly better, but it is still subjective and untestable, because "differentiated" is an extremely vague term. Putting numbers on it does not necessarily help; for example, if you refine your definition by saying that "a race is a group of organisms that are 50 percent divergent from the next most closely related group of things," you still have two problems. The first is that the 50 percent figure is entirely arbitrary, and others might think this "magic" number is not so magical. Most scientists will agree that genetic or morphological cohesion, reproductive isolation, or both lie at the core of what a species is. But there is no consensus as to what degree of divergence is significant as entities go their own separate ways in nature. For one group the magic number might be 5 percent, so that if it achieves over a 5 percent divergence level the probability is high that it will end up with complete divergence—hence becoming a new species. But for another group of organisms, the magic number might well be 95 percent.

A second problem is that whatever percent divergence you might choose, it must mean something biologically. The species definition that most taxonomists use requires 100 percent divergence in relevant traits. The question is an either/or one, and there is no subjectivity to it. The biological meaning of that 100 percent is that your entity is no longer meaningfully reproducing or significantly swapping genes with its closest relatives. It is on its own separate and historically established evolutionary trajectory. Percent divergence might mean something if researchers could pin down a magic threshold; but as we have seen, this is a very slippery concept indeed.

Yet this is how Wade described the process of species formation in a broadcast interview on CNN in June 2014: "Since evolution happens all the time, it's a continuous, unstoppable process that as a population splits, the two halves will continue to evolve, but now independently. So, over time they will accumulate differences between each other and eventually they'll become new species."

While we know from experience that radio interviews can be harrowing, and that it is difficult to explain things completely in short sound bites, this description of species formation is quite close to the portrayal he provides in his book. And what is particularly enlightening is that, directly before offering this definition, Wade had said: "Regionality underlines the fact of race because the populations on each continent have been evolving independently since we left our African homeland about 50,000 years ago."

The subjective perception of species, population evolution, and regionality expressed here leads to unwarranted conclusions about the existence of any entity below the level of the species *Homo sapiens*. This appears to reflect a failure on Wade's part to grasp the subtleties of taxonomic science. And this misapprehension has led to the third error we see in his reasoning.

3. A Misunderstanding of the Rigors of Taxonomic Science

Understanding our origins, and indeed the biology of all organisms on the planet, is really a problem of taxonomy. This vital branch of natural history is sometimes derided as "stamp collecting," but this claim could hardly be farther from the truth. Taxonomy is a well-developed and highly scientific endeavor that has been around in some form ever since humans began to name things. The science of taxonomy combines simple but rigorous hypothesis-testing approaches with objective definitions of species. It is true that competent taxonomists occasionally use the terms "subspecies" and "race" in their descriptions, but only as conveniences to imply future hypotheses to be tested.

Why is this "name game" so important for a discussion of race? We point to the following example from the pseudoscientific and white nationalist "journal" the *Occidental Quarterly*. This publication claims to have an academic and intellectual approach to the issues it addresses, but whiteness and nationalism almost always become the focus. The racist way the journal has approached the science of genetic and genomic bearings on race has been likened to some of the more extreme material that regularly appears in *Breitbart News*. In 2008, Andrew Hamilton (a pseudonym) published an article entitled "Taxonomic Approaches to Race." He started with a quote from the journalist Natalie Angier, who has written about Craig Venter, one of the scientists who led the release of the first draft of the human genome sequence. Angier characterized Venter—correctly—as implying that "there is only one race—the human race." Hamilton then continued by declaring, "The mawkish sentiment reported by Natalie Angier is incorrect. In biology, race is a synonym for subspecies" (Hamilton 2008, 11).

He went on to point out that race designations pervade the study of the natural world in organisms like insects, rabbits, cats, dogs, mice, horses, etc. But then he defined race as "a recognizably distinct division . . . distinguished from other groups by its unique clustering of genetically transmitted anatomical, physiological, psychological, and behavioral traits" (Hamilton 2008, 12). Well, thank you, Mr. Hamilton (or whatever your real name is) for clarifying that you feel race and subspecies are synonyms and that clusters are an important concept in naming races. But only one of these statements is correct (race equals subspecies); and much more importantly, the other (clusters are a test of taxonomic unit existence) is flat-out wrong, as we have seen.

Hamilton's declaration raises two important factors that we have pointed to throughout this book. Subspecies (races) are indeed recognized ranks; but as we hope to have shown, a subspecies rank is a placeholder, a hypothesis of species existence. Once this hypothesis is rejected, we are obliged to leave it and move on to other more testable and more interesting hypotheses. And in the human case, this hypothesis has repeatedly been tested—and rejected. We have discussed the clustering approach to organizing populations at length and have shown that it has severe shortcomings in testing hypotheses about the existence of taxonomic units. At best, the clusters are hypotheses that themselves need to be tested.

The genomic approach to the existence of races in human beings has usually involved collecting the frequencies of variants at many locations in the human genome from increasingly larger numbers of people. Of course, nobody would put much stock in a test of a hypothesis that involved only two individuals from each of the geographic regions suspected of diverging. For, as we have seen, if one examines too few individuals there is a danger of overdiagnosing the number of entities (i.e., of finding purely random evidence for differentiation). Similarly, examining too few populations will also result in overdiagnosis. Consider the following scenario: populations of a cosmopolitan organism are examined for their genetic variability by sequencing the genomes of individuals from Africa and Oceania. Not surprisingly, some genetic differences are detected and found to be significant, in that some are unique to the individuals from Africa while others are unique to individuals from Oceania. A big hoopla could be made, and species existence could be claimed; but this would be poor science, because the severity of the test is so low as to make it meaningless. Why? Because the organism might also exist in Europe, the Americas, and East Asia. By leaving out the populations "in between," one would miss the connectedness of the two populations initially sequenced. This phenomenon in widely distributed populations led the human biologist Frank Livingstone many years ago to conclude that "there are no human races, there are only clines" (Livingstone 1962, 279).

Wade understands this. Here is how he describes a genome-level polymorphism study and how it can be interpreted in a taxonomic context. He first uses the Rosenberg et al. (2002) study that we discussed in the last chapter to suggest that there are five clusters of people on the planet. As we saw, this important study used genomic information (nearly four hundred markers) from one thousand people, and employed the STRUCTURE clustering approach. The one thousand subjects "clustered naturally into five groups, corresponding to the five continental races." Wade apparently feels it is unimportant that this study was soon criticized by several researchers who objected that intermediate populations needed to be examined to exclude potential clinal variation. He then describes the Rosenberg group's next study of 2005, which increased the number of markers to nearly one thousand and yielded similar results (Ramachandran et al. 2005). And he suggests that the addition of more data in this case addressed the "cline" criticism. But while more data always help, the critical addition in this case would not be more genetic markers but more individuals from different geographic areas. These were not forthcoming, but Wade nevertheless uses the expanded genomic information (i.e., the doubling of the number of markers) to state categorically that "they found the clusters are *real*." [italics added].

More importantly for our argument about taxonomy, Wade goes on to discuss the inclusion of new information (using a newer genetic survey technology than the one used in the Rosenberg group's studies) to address the problem. In this newer study, one thousand different individuals were surveyed, but from fifty-one well-defined geographic areas. And instead of five, the researchers involved clustered their subjects into seven major groups. What is more, when even more subjects were added to Rosenberg and colleagues' data set, as was done by Sarah Tishkoff and her colleagues, fourteen clusters were inferred (Bryc et al. 2010). You might smell a rat here. But here is how Wade handles this new information:

> It might be reasonable to elevate the Indian and Middle Eastern groups (the two new ones) to the level of major races, making seven in all. But then many more subpopulations could be declared races, so to keep things simple, the five-race continent based scheme seems the most practical for most purposes. (Wade 2015, 177).

Any self-respecting taxonomist would avoid the kind of language Wade uses here. It is unscientific and circular. He is saying that just because inferences about the number of races vary, we cannot say that races do not exist. This is a variation on the argument that while people disagree on the number of shapes, shapes still exist—which simply trivializes the definitions we use in science generally and in taxonomy specifically.

There are six to seven billion human beings on the planet, and the best test of any hypothesis about human genomes and populations would include them all. Of course, this is not possible at present and probably never will be. But if it *were* possible, and the clustering were performed as in the two studies we have just referred to, we wonder how many groups might fall out. We suspect that, depending on the markers used, it might be as many as the entire number of nuclear families there are on the planet. Certainly, the patterns that would emerge from such a global analysis would not be anywhere near a clear demonstration of any definition of race that one could generate. Clustering alone is obviously inadequate for addressing problems of this kind in taxonomy and systematics. Which brings us to the fourth mistake made by proponents of a biological basis for race.

4. Misunderstanding the Meaning of Clustering and Evolutionary Trees

Wade's "evidence" for the biological basis of races is based purely on clustering. But clustering is only one way genetic data (or any other kind of discrete data) can be analyzed to test hypotheses. Perhaps a better approach is to use a branching diagram based on reconstructing the evolutionary events that led to the branches. Significantly, Wade does not present this kind of information or analysis in his book, possibly because researchers have for a long time realized that branching diagrams cannot represent the patterns of evolution of individuals that belong to the same species—something that directly reflects the difficulty and artificiality of sorting individuals into races. Branching diagrams can be very useful when used on single genes, and are extremely informative when used on clonal molecules like the maternally inherited mitochondrial DNA and the paternally inherited Y chromosome. But to our knowledge, no properly conceived attempt to build evolutionary trees with many recombining genetic regions, such as those on our autosomes, has ever resulted in a tree with any significant resolution.

The bottom line here, then, is that any attempt to arrive at a hierarchical structuring of humans using phylogenetic trees based on the entire genome will give us an unrecognizable and unresolved bush. But if that is the case, why do clustering methods appear to recover "structure"? We suggest that part of the reason relates to the next mistake in our list, the one that is made in doing genetic studies of geographically separated human populations by cherry-picking, or the phenomenon we prefer to call the "Stephen Colbert effect."

5. Cherry-picking AIMs: The Stephen Colbert Effect

Most of the early clustering studies used around one thousand genetic markers. More recent studies up the ante into the hundreds of thousands. These markers are chosen because they are believed to be informative about human ancestry, which as we have already noted is why they are known as "ancestral informative markers," or AIMs. These markers are established using what we like to call the "white swan" principle. People of different geographic origins have their genomes scanned, and when a variant appears at high frequency for a particular geographic location, that variant is said to be a marker for people from the geographic region concerned.

It is safe to say that this procedure introduces a bias into how the data are interpreted. This bias is so extreme that, when Stephen Colbert was presented with a genetic survey of his genome on the PBS show *Faces of America*, he was told he is 100 percent Caucasian. Some of the other guests were given similar results: Yo-Yo Ma was told he is 100 percent Asian. But some individuals were shocked by their results, among them Eva Longoria, who was given figures that deviated considerably from her prior view of her ancestry.

What was going on? Well, currently there are nearly thirty million places along the chromosomes of humans (out of three billion total) at which humans may vary. But any two randomly chosen humans will typically vary at only about three million bases. So if the typical ancestry study uses three hundred thousand markers (not too far from the actual number examined by commercial laboratories nowadays), it will only be looking at 10 percent of the potential differences between any two genomes, or about 0.1 percent of the entire genome. At best, then, these studies scan less than 1 percent of a human genome. What about the other 99 percent? Much of this remainder is not variable; but that part of it which is variable is African in origin. This means that 99 percent of the total variation in any human genome should be considered African. And what that in turn means is that Stephen Colbert is actually 99 percent African, and at most 1 percent Caucasian.

An argument commonly used to contradict this observation is known as the "Mount Everest paradox." The argument goes as follows: The elevation of Mount Everest differs from the surface of the ocean by an incredibly small fraction (about 0.0008) of the earth's diameter. But anyone standing at the foot of Mount Everest can tell the difference, and it is huge. Again, though, this is a trivial and unscientific argument. One could just as easily argue that, to a bacterium, a golf ball looks like Mount Everest (indeed, a similar 0.0002 percentage diameter-wise is involved). But any golfer can tell you how hard it is to find a golf ball in the rough. It is not the changes or differences that matter, but rather what the differences mean, and whether there is any objective way to interpret them. Some researchers prefer to interpret this information in the context of ancestry, which brings us to the sixth major mistake Wade makes.

6. Conflating Racially Based Genetic Differences with Explanation of Ancestry

The broad availability of genetic testing has made it something of a routine among people who are interested in their ancestry. But what do those ancestry tests actually tell us? They basically inform us about the chunks of DNA in our genomes and where that DNA might have originated. This is, of course, about individuals, and more specifically about genes; but very unfortunately, ancestry testing has become a proxy for race determination. This is particularly unfortunate, because everyone's genome is a mosaic of ancestries, even including chunks of DNA that might show ancestry from other species. And it is particularly misleading when it comes to our understanding of race in humans, because there are no definitions as to how many genomic variants (and even more complicated, *which* variants) might make a difference. Because ancestry can be traced unbroken all the way to the nuclear-family level, we strongly suggest that the ancestry approach is not informative in any way relative to the hypotheses we proposed in the first part of this piece. Like race, ancestry is clinal with respect to any purported higher level and simply connects us with one another. So what, in the end, do those genetic ancestry tests tell us? Perhaps the best way to view the whole ancestry business is to use a term that has recently appeared in the literature to describe commercially available ancestry tests: "recreational genomics." Such recreational approaches offer little, if anything, to science. It is even arguable whether they offer much to those engaged in the recreation.

It is often argued that we need to speak about races if we wish to study the movements of people around the planet and determine their evolutionary histories. But this is entirely false, because our knowledge of mtDNA and Y chromosomes, supplemented by the archaeological and fossil records, already provides an excellent picture of how humans migrated in the past. We are not impeded at all in these endeavors by the lack of formally defined biological races, because we use clinal markers that follow individual haplotypes. No a priori definition of race is needed to interpret the results of such tree-based analyses. It is also argued that ancestry is an important component in medicine; and the leap is then made to the claim that knowledge of a person's race is essential to the maintenance of his or her health. However, medicine is already benefiting from individualized genomics; and because, as we pointed out in our book *Race? Debunking a Scientific Myth*, race and ancestry have in the past proven poor clinical tools, we see few cogent reasons to consider race a major factor in medicine. Perhaps ancestry will prove to be important; but the concept of race in medicine is clearly barking up the wrong tree.

7. Conflating Variation and Allele Frequency Differences with Adaptation (and Hence Elements of the Human Condition)

Adaptation and allele frequencies are the focus of Wade's last five chapters and were extensively discussed in our *Race? Debunking a Scientific Myth*. Wade seems to believe that we need a notion of races to explain why some of us look different from others. Yet nearly all the (remarkably few) "adaptations" that can be identified—for example, the diverse responses to high-altitude living and to living under intense solar radiation—tend to be intensely local in their occurrence and are not at all usefully illuminated by any concept of major "races."

Charles Murray placed his bet—that attacks on Wade's book would be made along more sociological than scientific lines—based on his allegation that scientists would fear breaking away from tyrannical orthodoxy. Wade addresses this tyranny issue in the first few pages of *A Troublesome Inheritance*. He is concerned that unorthodox thinking tends to get stifled by convention such that progress, both scientific and social, is impeded. We could not disagree more with Murray and Wade on this matter, but in this book we have deliberately refrained from going anywhere near that kind of argument. To us, the most important thing is that when the science Wade cites is placed under real scrutiny, the thesis of his book fails miserably, and Mr. Murray loses his bet.

Wade's insistence that science advances by departure from orthodoxy might be called the "Indiana Jones fallacy." And it is especially important to understand quite how destructive this fallacy is, because in line with what appears to be a growing distrust of science in many corners of society, all the authors of positive reviews of Wade's book (Murray's included) have harped on the far-reaching importance of Wade's own departure from despotic scientific orthodoxy. As scientists, we recognize how gratifying it would be if every published scientific paper was earthshaking and unorthodox. This would make scientific progress excitingly rapid and unbounded. But the sad truth is that much of science is rather boring and procedural—just as rigor demands.

Even the hypothesis that there are genetic differences among people from different geographic regions—classifiable or not—is entirely mundane. Of course there are differences among us, as there are among members of any widespread species. We do not need to spend millions of dollars sequencing genomes to know this. The real questions are whether the differences are in any way significant; whether they are interpretable in a rigorous scientific context; and whether the classification of people into races does anything useful to help us understand them. The answers here are clear. While the differences are there, they are superficial—for the most part they are simply epiphenomena of the last few tens of millennia of human evolution—and they do not sort out well at all on larger scales. And, as to whether racial classifications help us to understand anything at all, the answer is a resounding "No!"

This last point may seem at first glance a bit counterintuitive, because as we noted at the beginning of this book, it is often possible to visually sort a certain proportion of your fellow citizens by general geographic origin. And indeed, for almost all the past fifty thousand years or so during which *Homo sapiens* has been widely present throughout the Old World, our hunting-gathering precursors were sparsely spread out across vast landscapes and constantly buffeted by rapidly changing climatic and environmental conditions. This provided optimal circumstances for regional differentiation and the incorporation of minor genetic novelties into local populations; and it explains why, for example, Africans generally tend to resemble each other more closely than they do Eastern Asians or Europeans. But throughout, all of us remained members of one single, interbreeding species; and we guarantee that the edges between populations were never sharp. What is more, over the ten thousand years that have elapsed since the adoption of a more settled way of life, demographic circumstances have changed entirely, as populations have mingled on large and small scales and often over vast distances. This, above all, is why it is hopeless to look for the boundaries that are necessarily there if we are to usefully recognize races or to pretend that the enterprise of looking for them is in any way scientific. The central tendencies may be there, but the boundaries are not. Which, of course, means that race is a totally inadequate way of characterizing diverse humankind, or even of helping us to understand humanity's glorious variety.

Epilogue

RACE AND SOCIETY

Up to this point we have been mostly concerned with what races most emphatically are *not*: meaningful units into which members of the species *Homo sapiens* can be grouped. Still, for reasons we have sporadically touched upon, it is useless to deny that to many people the idea of race *feels* real; and indeed, in most human societies it is a concept that unconsciously or otherwise pervades people's experience, mediated by ways of thinking about human variety and interaction that are absorbed remarkably early in life. Those social attitudes are often reinforced by the penchants of governments to classify their citizens along racial lines.

An excellent example of how racial attitudes can insidiously permeate a society, even one in which "affirmative action" policies are widely implemented to address inequality among social groups, is provided by a recent analysis of the extensive opposition to paying college athletes, the only participants in a hugely lucrative industry who do not get lavishly remunerated. True, top college athletes receive scholarships to cover the costs of their education—although the rigor of that education is sometimes compromised in the interests of maximizing their availability to generate income for the institutions for which they play. But even under the best of circumstances, the financial value of the education the athletes receive is peanuts compared with the huge sums that colleges rake in from sports. In college football alone, and merely for TV rights to broadcast the seven annual playoff games, ESPN is contracted to pay the National College Athletic Association (NCAA) $7.3 billion from 2014 to 2026.

Rewarding top student athletes with scholarships is clearly the equivalent of paying star actors chump change for appearing in blockbuster movies; and this striking departure from standard ideas of fairness is normally justified by resorting to the "special status" of amateur sport, in which supposedly it is not winning or losing but the game itself that is important. According to the NCAA, "one of the biggest reasons fans like college sports is that they believe the athletes are really students who play for the love of the sport." Such self-serving rationalizations have deservedly attracted scorn from commentators. One of them, the author Taylor Branch, wrote in the *Atlantic* that "two of the noble principles on which the NCAA justifies its existence—'amateurism' and the 'student-athlete'—are cynical hoaxes, legalistic confections propagated by the universities so that they can exploit the skills and fame of young athletes" (Branch 2015).

While the motives of the NCAA and the colleges in all of this are transparent, those of the audience are more opaque. Accordingly, a group led by the political scientist Kevin Wallsten of California State University–Long Beach recently looked in detail at the fan base to which the NCAA typically appeals to justify its inequitable position. True, a 2015 Marist poll had found that a whacking 65 percent of respondents overall opposed paying college football and basketball players, a group among which African Americans are disproportionately represented. But when Wallsten and colleagues broke down respondents in the 2014 Cooperative Congressional Election Study (CCES) by race, the researchers discovered that those who identified themselves as black were significantly more likely than self-identified whites to support payment to those college athletes. More than half of the blacks surveyed wanted them to be paid, in contrast to fewer than a quarter of the whites.

Digging deeper, the researchers undertook a survey of the 2014 CCES sample specifically to discover whether prejudiced racial views influenced the white respondents' antipathy toward paying college athletes. And, disturbingly, they reported that the more negatively white respondents felt about blacks, the more vehemently they opposed paying the athletes. To double-check this result, some white respondents were shown pictures of young black men before being asked their feelings about paying college athletes, while others were not. Tellingly, those who had been shown the pictures were significantly more likely to say they opposed payment; and the strength of that opposition scaled with the degree of their expressed resentment of blacks.

On the strength of this example alone, America clearly still carries a damaging burden of racial resentment and prejudice. The days of overt lynchings may be over, but those of hugely disproportionate incarcerations of disadvantaged African Americans most certainly are not; and even such ostensibly race-neutral topics as whether or not college athletes should be paid for their efforts are evidently deeply influenced by racial feelings.

This makes it very plain that, although stratifying respondents by race in the college sports study might have been a biologically meaningless exercise, in cultural terms it was not. Human beings are, and always have been, intensely social creatures, and as such they crave group identity. The evolutionary psychologist Robin Dunbar has pointed out that in practical terms the maximum number of close associates the average person can keep close track of is a mere 150 or so; and although our modern urban society requires that we need to interact closely with significantly larger numbers of individuals over time, there is certainly a limit to the number of people we can directly relate to on a personal level (although there always seems to be room for one more).

In any complex human society, individuals need to identify with larger groups than the ones composed of the people they know and can identify with directly; and the only way to define groups of that kind is not as the sum of the individuals who compose them, but more abstractly, via some perceived property: nationality group, religious group, interest group, social stratum, and so on. And of all such categories, the racial grouping is intuitively among the most obvious, because it is signaled by physical appearance and can be perceived at a distance. Until we look more closely, of course.

We have already discussed at length the biological fallacy inherent in trying to classify people into racial groups; and now in closing we need to draw attention once again to the substantial disconnect that exists between perceived race and culture. Within a given society, racial self-attribution might well broadly indicate an individual's affinity with a specific cultural group. But identity is not a biological attribute; and indeed, in a multiethnic society it is something an individual often merely assumes. What is more, if it is the cultural entity that is of interest, it is surely much more accurate and meaningful to denote and characterize it in cultural terms.

Across societies and continents, of course, the correlation between race and culture breaks down entirely. Knowing how an individual is racially classified doesn't tell you much about him or her as an individual or even what social norms he or she espouses. A "black" person from Jackson and another from Juba or Jaffna will find little culturally or linguistically in common; and it is impossible from anyone's racial classification, whether self-assigned or governmentally imposed, to predict such fundamental individual qualities as his or her personality type, brainpower, or talents.

Caution is also in order, because racial classification is something of a self-fulfilling prophecy that allows us to dig ourselves ever deeper into a self-sustaining cycle. For when we attempt to deconstruct the population into racial categories, even for the most benign of purposes such as the remedying of social inequities, we also find ourselves reinforcing those categories—and, incidentally, the entire litany of stereotypes that inevitably comes along with them. "Us versus them" lies at the heart of all internecine conflicts. It is surely much better to recognize immediately that breaking the population into supposed biological groupings helps perpetuate the divisions among those assumed groups. If we wish to indulge in social engineering, we would plainly be much better off identifying the cultural or economic groups we wish to habilitate and defining them in the appropriate terms.

But there remains that awkward matter of that innate human need to name and classify everything around us. This process of atomization and categorization is, after all, fundamental to language; and it is language that provides the armature for our intellectual understanding of the world around us, including our own species. Language demands that we give things names; and while people invariably exist along a spectrum of variation in any behavioral or physical quality you might want to specify, names are hard-and-fast categories. Even when we attach modifiers, the very act of naming presupposes archetypes. The bottom line is that when we ascribe a limited number of names to elements lying along a spectrum, we create a set of artificial and arbitrary categories; and if the underlying entities are multidimensional ones, varying in more respects than, say, the colors of the visible spectrum, the distortions only worsen as each new consideration is added—as, of course, inevitably happens when we carve up the gloriously heterogeneous human population into discrete "races."

Take, for example, the doomed efforts of well-meaning biological anthropologists of the immediate post–World War II period to reclassify variation within our species *Homo sapiens*. With the racial atrocities of the 1930s and 1940s only too fresh in their minds, they were painfully aware of the dangers and difficulties attendant on biologically classifying people; but many were not yet able to extricate themselves from the feeling that, since variation was evidently there, scientists should nonetheless classify it. This was clearly the point of view of the thoughtful developmental anthropologist Stanley Garn, who attempted a new taxonomy of *Homo sapiens* in 1961. Having settled, after much agonizing, on nine major "geographic races," Garn rapidly found himself bogged down in a morass of "local races" within each of these larger units; and he ultimately found it necessary to concede that each local race contained innumerable "micro-races." Fortunately, he ran out of steam well before the potential infinite extension of this logic forced him to conclude that every individual belonged to his or her own unique race.

We mention Garn's experience merely to illustrate the point that the deformation of fact introduced by the attempt to divide up a heterogeneous, complex, and dynamically reticulating entity such as *Homo sapiens* can only be eliminated by totally abandoning the effort to subdivide. The resulting categories cannot simply be fine-tuned to eliminate distortion. Anthropologists of Garn's generation bore a burden of erroneous received wisdom that it proved difficult for many to shed, even after the experience of living through the darkest racial episode of recent history. But perhaps, going forward, the ultimate abandonment of scientific and official racial categorization will be made easier by the explicit realization that, certainly from a biological point of view, doing so will cost us absolutely nothing, intellectually, scientifically, or financially. After all, it is becoming increasingly evident that making racial distinctions within the human species not only adds no useful heuristic to our understanding of human variety, but makes any useful attempt to understand it harder.

All well and good; but concepts of race, explicit or otherwise, are clearly and tenaciously entrenched in almost everyone's consciousness. And they still lie at the root of some of the most distressing and stubborn social problems that exist around the world. How can we eliminate them—or at least attenuate them—in favor of understanding variation within our species in more constructive and useful ways? Well, one size will clearly never fit all in a problem this vast. Racial attitudes are embedded in such a variety of different narratives, in such a multiplicity of cultures and contexts across the planet, that there will never be a single solution to the problem of easing the evils associated with the social perception of race. Governments can help to move the dial by passing legislation designed to guide and modify social attitudes over the long term; but a genuinely enduring commitment to clearly visualized goals is invariably required, something that it is often hard to envisage. Still, in some places there do appear to be modest grounds for optimism, at least for those prepared to take a long view.

One of those places is the United States, where in 2013 the sociologists Tatishe Nteta and Jill Greenlee used the 2008 election of Barack Obama to the presidency to test the "impressionable years" hypothesis, which proposes that political events experienced in early youth can have lasting effects on attitudes held in later life. Previous studies had suggested that, among American whites born since the 1970s, long-held beliefs in the biological inferiority of American blacks had been consistently waning, and that Obama's ascent to the presidency had accelerated this decline. Using data from the 2010 CCES survey, Nteta and Greenlee approached the issue from a slightly different perspective, asking not whether Obama's election could be viewed as a product of this existing trend, but rather if it had the potential to catalyze further change in social attitudes.

Nteta and Greenlee's first finding was that an "Obama generation" of those whites who fell into the critical 18–25 "young adult" age bracket during Obama's presidential campaign could indeed be identified, via racial attitudes that were consistently more liberal than typical of earlier generations. Looking more closely, though, they additionally discovered that higher levels of education in this generation did not, as in its predecessors, correlate with reduced levels of racial resentment (as opposed to "old-fashioned racism," which involves a more general belief in inferiority).

They examined two possible explanations for this. One was that, by promoting individualistic beliefs, education may simultaneously enhance negative views of perceived work ethic. The other was that "if social norms regarding racial equality are increasingly widespread, the role of educational institutions in promoting racial equality may be less important" (Nteta and Greenlee 2013, 892). The authors leaned toward the more optimistic latter conclusion; but we cannot help observing that the issue would not exist were it not for the prior categorization of the populations concerned.

Nteta and Greenlee are frank about the difficulties involved in predicting the future of what is a very complex social dynamic. And, of course, their research did not touch on any potential evolution in the attitudes of black Americans toward white ones. But they end their study on a hopeful note, more encouraged by the apparent waning of "old-fashioned racism" than discouraged by the reverse effect of education on racial resentment—a complex of attitudes that will always be much more greatly affected by prevailing economic circumstances than by anything else. Nteta (website) has since observed: "The ascendance and eventual election of Obama may have led to the formation of a new generational grouping, and . . . this generation's racial attitudes represent the culmination of the nation's steady march toward racial reconciliation and equality." Whether there will be good reason to maintain such optimism over the life of the succeeding administration remains to be seen; but the auguries are not necessarily good. In February 2017, the *New York Times* reported that Stephen K. Bannon, then President Donald Trump's chief strategist, was versed in the works of Julius Evola, an Italian philosopher who considered the fascists "overly tame" on matters of race, and who has recently become a darling of the emerging "alt-right." As science, race may (or should) be a dead issue; but it shows zombie-like tenacity on the social and political fronts.

Notes and Bibliography

In the following section we list by chapter all titles mentioned in the text, including all sources of quotations and figures.

Preface

This book is a complement to our previously published work on race, *Race? Debunking a Scientific Myth* (Tattersall and DeSalle 2011). Anyone interested in the history of race in Western thought, in aspects of biological adaptation, or in race in the context of the larger story of human evolution is encouraged to look at this volume as well. Since it was published, our colleague Bob Sussman entered the fray with his *The Myth of Race*, a compelling and unsettling examination of the interface of race "science" with ideology and politics. Readers with an interest in the eugenics movement are particularly recommended to consult this important work. Finally, we reference Nicholas Wade's *A Troublesome Inheritance* here, if only to point to an excellent example of why books such as the present one are still so badly needed.

Sussman, R. W. 2014. *The Myth of Race: The Troubling Persistence of an Unscientific Idea*. Cambridge, MA: Harvard University Press.
Tattersall, I., and R. DeSalle. 2011. *Race? Debunking a Scientific Myth*. College Station: Texas A&M University Press.
Wade, N. 2015. *A Troublesome Inheritance: Genes, Race and Human History*. New York: Penguin.

1. Evolutionary Lessons

Charles Darwin's notebooks can be accessed online at the website Darwin Online (http://darwin-online.org.uk/manuscripts.html). Darwin's (1859) *On the Origin of*

Species is easily accessed from the reference below or online at https://en.wikisource .org/wiki/On_the_Origin_of_Species_ (1859). *The Descent of Man* is equally widely available. Histories of early evolutionary thought can be found in books by two prominent evolutionary biologists—Ernst Mayr (1982) and Stephen J. Gould (2002). Thomas Malthus's (1798) short treatment of populations is in his *An Essay on the Principle of Population,* available online at https://archive.org/details/principleessay-on00maltrich. Populational versus typological thinking is discussed by Mayr (1984). The wonderful spandrels analogy was first published by Gould and Lewontin (1979).

Darwin, C. R. 1859. *On the Origin of Species by Means of Natural Selection; or, The Preservation of Favoured Races in the Struggle for Life.* London: John Murray.

——. 1871. *The Descent of Man.* 2 vols. London: John Murray

Gould, S. J. 2002. *The Structure of Evolutionary Theory.* Cambridge, MA: Harvard University Press.

Gould, S. J., and R. C. Lewontin. 1979. "The Spandrels of San Marco and the Panglossian Paradigm: A Critique of the Adaptationist Programme," *Proceedings of the Royal Society of London B: Biological Sciences* 205 (1161): 581–598.

Malthus, T. R. 1798. *An Essay on the Principle of Population; or, A View of Its Past and Present Effects on Human Happiness.* London: John Murray.

Mayr, E. 1982. *The Growth of Biological Thought: Diversity, Evolution, and Inheritance.* Cambridge, MA: Harvard University Press.

——. 1984. "Typological Versus Population Thinking." In *Conceptual Issues in Evolutionary Biology,* ed. E. Sober, 14–38. Cambridge, MA: MIT Press.

2. Species and How to Recognize Them

We discuss DNA biology in detail in the companion volume to the present book (Tattersall and DeSalle 2011). Mayr's species concept is published in many of his books, but a concise summary of the concept is given in an excellent paper by Kevin de Queiroz (2005), who also discusses his own ideas about species. Ghiselin (1974) presents the author's "radical solution to the species problem." Linnaeus's four-race system is discussed in detail by Gould (1994).

Ghiselin, M. T. 1974. "A Radical Solution to the Species Problem." *Systematic Biology* 23 (4): 536–544.

Gould, S. J. 1994. "The Geometer of Race." *Discover* 15 (11): 65–69.

Haeckel, E. 1879. *Evolutionary History of Man.* New York: Appleton.

Queiroz, K. de. 2005. "Ernst Mayr and the Modern Concept of Species." *Proceedings of the National Academy of Sciences of the United States of America* 102 (suppl. 1): 6600–6607.

Tattersall, I., and R. DeSalle. 2011. *Race? Debunking a Scientific Myth.* College Station: Texas A&M University Press.

Watson, J. D., and F. H. C. Crick. 1953. "A Structure for Deoxyribose Nucleic Acid." *Nature* 171: 737–738.

3. Phylogenetic Trees

Lamarck's *Philosophie zoologique* (1809) can be found online at www.ucl.ac.uk/taxome/jim/Mim/lamarck_contents.html. For an interesting discussion of the phylogenetic tree battles, see Hull (2010). Sneath and Sokal's seminal work is described in their book *Numerical Taxonomy* (1963), and Hennig's treatise on systematics was translated into English in 1966 (Hennig and Davis, 1966). Edwards and Cavalli-Sforza outlined their likelihood approach in a paper in *Heredity* (1963). For a general but thorough treatment of phylogenetic approaches, see DeSalle and Rosenfeld's (2013) *Primer of Phylogenomics*.

DeSalle, R., and J. Rosenfeld. 2013. *Phylogenomics: A Primer*. New York: Garland Science.

Edwards, A. W. F., and L. L. Cavalli-Sforza. 1963. "The Reconstruction of Evolution." *Heredity* 18: 553–554.

Hennig, W., and D. D. Davis. 1966. *Phylogenetic Systematics*. Champaign-Urbana: University of Illinois Press.

Hull, D. L. 2010. *Science as a Process: An Evolutionary Account of the Social and Conceptual Development of Science*. Chicago: University of Chicago Press.

Lamarck, J.-B. 1809. *Philosophie zoologique*. Paris: Deterville.

Sneath, P. H. A., and R. R. Sokal. 1963. *Numerical Taxonomy. The Principles and Practice of Numerical Classification*. San Francisco: Freeman.

4. The Name Game: Modern Zoological Nomenclature and the Rules of Naming Things

An excellent summary of taxonomic practice, development, and history can be found in Yoon (2010). Mayr's definitions were first articulated in his text *Principles of Systematic Zoology*, first published in 1969. William Stearns's original designation of Linnaeus as the type of *Homo sapiens* appeared in 1959. Notton and Stringer (2014) also discuss the status of the type specimen of our species. Reproductions of *Systema Naturae* and *Regnum Animalia* can easily be found online, as can the nomenclatural rules for animals (www.nhm.ac.uk/hosted-sites/iczn/code/index.jsp?booksection=introduction&nfv=) and plants, fungi, and algae (www.iapt-taxon.org/nomen/main.php).

Mayr, E. 1969. *Principles of Systematic Zoology*. New York: McGraw-Hill.

Notton, D., and C. Stringer. 2014. "Who Is the Type of *Homo sapiens*?" International Commission on Zoological Nomenclature website. http://iczn. org/content/who-type-homo-sapiens, accessed January 15, 2014.

Stearns, W. T. 1959. "The Background of Linnaeus's Contributions to the Nomenclature and Methods of Systematic Biology." *Systematic Zoology* 8 (1): 4–22.

Yoon, C. K. 2010. *Naming Nature: The Clash Between Instinct and Science*. New York: Norton.

5. DNA Fingerprinting and Barcoding

The identification of Canada geese as the cause of the crash of Flight 1549 is documented in Marra et al. (2009). Paul Hebert's original discussion of DNA barcoding is in Hebert, Cywinska, and Ball (2003), and his collaborative work with Dan Janzen and Winnie Hallwachs is reported in Hebert et al. (2004). The BoLD databases are online at www.boldsystems.org/index.php/databases. CODIS is also online at www .fbi.gov/services/laboratory/biometric-analysis/codis. How microsatellite data are apportioned in populations is discussed by Budowle et al. (2001) and Willems et al. (2014). Integrated taxonomy is discussed in detail by DeSalle, Egan, and Siddall (2005) and Miller (2007).

Budowle, B., B. Shea, S. Niezgoda, and R. Chakraborty. 2001. "CODIS STR Loci Data from 41 Sample Populations." *Journal of Forensic Science* 46 (3): 453–489.

DeSalle, R., M. G. Egan, and M. Siddall. 2005. "The Unholy Trinity: Taxonomy, Species Delimitation and DNA Barcoding." *Philosophical Transactions of the Royal Society B: Biological Sciences* 360 (1462): 1905–1916.

Hebert, P. D. N., A. Cywinska, and S. L. Ball. 2003. "Biological Identifications Through DNA Barcodes." *Proceedings of the Royal Society of London B: Biological Sciences* 270 (13): 313–321.

Hebert, P. D. N., E. H. Penton, J. M. Burns, D. H. Janzen, and W. Hallwachs. 2004. "Ten Species in One: DNA Barcoding Reveals Cryptic Species in the Neotropical Skipper Butterfly *Astraptes fulgerator*." *Proceedings of the National Academy of Sciences of the United States of America* 101 (41): 14812–14817.

Marra, P. P., C. J. Dove, R. Dolbeer, N. F. Dahlan, M. Heacker, J. F. Whatton, N. E. Diggs, C. France, and G. A. Henkes. 2009. "Migratory Canada Geese Cause Crash of US Airways Flight 1549." *Frontiers in Ecology and the Environment* 7 (6): 297–301.

Miller, S. E. 2007. "DNA Barcoding and the Renaissance of Taxonomy." *Proceedings of the National Academy of Sciences of the United States of America* 104 (12): 4775–4776.

Willems, T., M. Gymrek, G. Highnam, D. Mittelman, Y. Erlich, and 1000 Genomes Project Consortium. 2014. "The Landscape of Human STR Variation." *Genome Research* 24 (11): 1894–1904.

6. Early Biological Notions of Human Divergence

Early studies of the enigmatic Kennewick Man skeleton are summarized by Chatters (2002). The most comprehensive recent account is by Burke et al. (2016), and Thomas (2000) elegantly placed the controversy surrounding it in its political context. An earlier take on the Morton/Gould affair was published by Michael (1988), with the most recent account written by Lewis et al. (2011). Jablonksi's (2013) book on skin color is the most recent comprehensive treatment of this topic. Lamason et al. (2005) worked out the dynamics of SLC24A5 involvement in skin pigmentation in zebrafish and humans. Morrison and colleagues' *The Genealogical World of Phylogenetic Networks* is online at http://phylonetworks.blogspot.com. Rieppel's (2010) paper on trees is a good starting place for trees and networks.

Burke, H., C. E. Smith, D. Lippert, J. E. Watkins, and L. J. Zimmerman, eds. 2016. *Kennewick Man: Perspectives on the Ancient One*. New York: Routledge.

Chatters, J. C. 2002. *Ancient Encounters: Kennewick Man and the First Americans*. New York: Simon and Schuster.

Gallet, F. 1800. *Arbre généalogique des langues mortes et vivantes*. Broadside sheet, privately published.

Hooton, E. A. 1946. *Up from the Ape*. New York: Macmillan.

Jablonski, N. G. 2013. *Skin: A Natural History*. Berkeley: University of California Press.

Keith, A. 1915. *The Antiquity of Man*. London: Williams and Norgate.

Lamason, R. L., M.-A. P. K. Mohideen, J. R. Mest, A. C. Wong, H. L. Norton, M. C. Aros, M. J. Jurynec, et al. 2005. "SLC24A5, A Putative Cation Exchanger, Affects Pigmentation in Zebrafish and Humans." *Science* 310 (5755): 1782–1786.

Lewis, J. E., D. DeGusta, M. R. Meyer, J. M. Monge, A. E. Mann, and R. L. Holloway. 2011. "The Mismeasure of Science: Stephen Jay Gould Versus Samuel George Morton on Skulls and Bias." *PLoS Biol* 9 (6): e1001071.

Michael, J. S. 1988. "A New Look at Morton's Craniological Research." *Current Anthropology* 29: 349–364.

Rieppel, O. 2010. "The Series, the Network, and the Tree: Changing Metaphors of Order in Nature." *Biology & Philosophy* 25 (4): 475–496.

Thomas, D. H. 2000. *Skull Wars: Kennewick Man, Archaeology, and the Battle for Native American Identity*. New York: Basic Books.

Weidenreich, F. 1947. "Facts and Speculations Concerning the Origin of *Homo sapiens*." *American Anthropologist* 49 (2): 187–203.

7. Mitochondrial Eve and Y-Chromosome Adam

The Hirschfelds' blood group work was published in a 1919 issue of *Lancet*, and Landsteiner's seminal research, which won him a Nobel Prize, is summarized in *The Specificity of Serological Reactions* (1936). Provine (2001) described the development of population genetics. Edwards and Cavalli-Sforza's first tree can be found in their paper published in 1963, and its subsequent revision appears in Cavalli-Sforza, Menozzi, and Piazza (1994). Lewontin's work was published in 1974. Mitochondrial Eve was first introduced in Cann, Stoneking, and Wilson (1987), and Van Oven and Kayser's global mtDNA tree appeared in 2009. The MITOMAP website summarizing the tree can be found at www.mitomap.org/foswiki/bin/view/MITOMAP/GBFreqInfo. The Hammer and Cavalli-Sforza Y-chromosome Adam work is discussed by Mitchell and Hammer (1996) and Ayala et al. (1995). Poznik and colleagues published the most recent and comprehensive Y-chromosome tree in 2016.

Ayala, F. J., L. L. Cavalli-Sforza, P. Menozzi, and A. Piazza. 1995. "Adam, Eve, and Other Ancestors: A Story of Human Origins Told by Genes." *History and Philosophy of the Life Sciences* 17 (2): 303–313.

Cann, R. L., M. Stoneking, and A. C. Wilson. 1987. "Mitochondrial DNA and Human Evolution." *Nature* 325 (6099): 31–36.

Cavalli-Sforza, L. L., P. Menozzi, and A. Piazza. 1994. *The History and Geography of Human Genes.* Princeton, N.J.: Princeton University Press.

Edwards, A. W. F., and L. Cavalli-Sforza. 1963. "The Reconstruction of Evolution." *Heredity* 18: 553–554.

Hirschfeld, L., and H. Hirschfeld. 1919. "Serological Differences Between the Blood of Different Races: The Result of Researches on the Macedonian Front." *Lancet* 194: 675–679.

Landsteiner, K. 1936. *The Specificity of Serological Reactions.* Springfield, IL: Thomas.

Leslie, S., B. Winney, G. Hellenthal, D. Davison, et al. 2015. "The Fine-Scale Genetic Structure of the British Population." *Nature* 519: 309–314.

Lewontin, Richard C. 1974. *The Genetic Basis of Evolutionary Change.* New York: Columbia University Press.

Mitchell, R. J., and M. F. Hammer. 1996. "Human Evolution and the Y Chromosome." *Current Opinion in Genetics & Development* 6 (6): 737–742.

Poznik, G. D., Y. Xue, F. L. Mendez, T. F. Willems, A. Massaia, M. A. Wilson Sayres, Q. Ayub, et al. 2016. "Punctuated Bursts in Human Male Demography Inferred from 1,244 Worldwide Y-Chromosome Sequences." *Nature Genetics* 48 (6): 593–599.

Provine, W. B. 2001. *The Origins of Theoretical Population Genetics, with a New Afterword.* Chicago: University of Chicago Press.

Tattersall, I. and R. DeSalle. 2012. *Race? Debunking a Scientific Myth.* College Station: Texas A&M Press.

Van Oven, M., and M. Kayser. 2009. "Updated Comprehensive Phylogenetic Tree of Global Human Mitochondrial DNA Variation." *Human Mutation* 30 (2): E386–E394.

8. The Other 99 Percent of the Genome

The human diversity array is discussed in detail by Lazaridis et al. (2014) and Patterson et al. (2012). Lachance and Tishkoff (2013) and McTavish and Hillis (2015) are good starting places for discussion of ascertainment bias. John Novembre and colleagues' work on European genetic structure can be found in Novembre et al. (2008).

Lachance, J., and S. A. Tishkoff. 2013. "SNP Ascertainment Bias in Population Genetic Analyses: Why It Is Important, and How to Correct It." *BioEssays* 35 (9): 780–786.

Lazaridis, I., N. Patterson, A. Mittnik, G. Renaud, S. Mallick, K. Kirsanow, P. H. Sudmant, et al. 2014. "Ancient Human Genomes Suggest Three Ancestral Populations for Present-Day Europeans." *Nature* 513 (7518): 409–413.

McTavish, E. J., and D. M. Hillis. 2015. "How Do SNP Ascertainment Schemes and Population Demographics Affect Inferences About Population History?" *BMC Genomics* 16 (1): 266.

Novembre, J., T. Johnson, K. Bryc, Z. Kutalik, A. R. Boyko, A. Auton, A. Indap, et al. 2008. "Genes Mirror Geography Within Europe." *Nature* 456 (7218): 98–101.

Patterson, N., P. Moorjani, Y. Luo, S. Mallick, N. Rohland, Y. Zhan, T. Genschoreck, T. Webster, and D. Reich. 2012. "Ancient Admixture in Human History." *Genetics* 192 (3): 1065–1093.

9. ABBA/BABA and the Genomes of Our Ancient Relatives

Meyer et al. (2014) report on the genome sequence of the hominin from Sima de los Huesos. The spectacular sediment work in Pääbo's lab is published in Slon et al. (2017). A clear summary of paleogenomes up to 2016 can be found in Slatkin and Racimo (2016). Krause and Pääbo (2016) have reviewed the state of the art. Martin, Davey, and Jiggins (2014), Green et al. (2010), Eriksson and Manica (2012), and Amos (2016) discuss at length the ABBA/BABA test of Neanderthal and Denisova introgression.

Amos, W. 2016. "The quantity of Neanderthal DNA in modern humans: A reanalysis relaxing the assumption of constant mutation rate." bioArxiv doi: http://doi .org.10/10.1101/065359.

Eriksson, A., and A. Manica. 2012. "Effect of Ancient Population Structure on the Degree of Polymorphism Shared Between Modern Human Populations and Ancient Hominins." *Proceedings of the National Academy of Sciences of the United States of America* 109: 13956–13960.

Ermini, L., C. Der Sarkissian, E. Willerslev and L. Orlando. 2015. "Major Transitions in Human Evolution Revisited: A Tribute to Ancient DNA." *Journal of Human Evolution* 79: 4–20.

Green, R. E., J. Krause, A. W. Briggs, T. Maricic, U. Stenzel, M. Kircher, N. Patterson, H. Li, W. Zhai, M. N. Fritz, et al. 2010. "A Draft Sequence of the Neandertal Genome." *Science* 328: 710–722.

Krause, J., and S. Pääbo. 2016. "Genetic Time Travel." *Genetics* 203 (1): 9–12.

Martin, S. H., J. W. Davey, and C. D. Jiggins. 2014. "Evaluating the Use of ABBA–BABA Statistics to Locate Introgressed Loci." *Molecular Biology and Evolution* 32 (1): 244–257.

Meyer, M., Q. Fu, A. Aximu-Petri, I. Glocke, B. Nickel, J.-L. Arsuaga, I. Martínez, et al. 2014. "A Mitochondrial Genome Sequence of a Hominin from Sima de los Huesos." *Nature* 505 (7483): 403–406.

Skoglund, P., D. H. Northoff, M. V. Shunkov, A. P. Derevianko, S. Pääbo, J. Krause, and M. Jakobsson. 2014. "Separating Endogenous Ancient DNA from Modern Day Contamination in a Siberian Neandertal." *Proceedings of the National Academy of Sciences of the United States of America* 107: 2229–2234.

Slatkin, M., and F. Racimo. 2016. "Ancient DNA and Human History." *Proceedings of the National Academy of Sciences of the United States of America* 113: 6380–6387.

Slon, V., C. Hopfe, C. L. Weiss, F. Mafessoni, M. de la Rasilla, C. Lalueza-Fox, A. Rosas, et al. 2017. "Neandertal and Denisovan DNA from Pleistocene Sediments." *Science* 356 (6338): 605–608.

10. Human Migration and Neolithic Genomes

Slatkin and Racimo summarize the paleogenomic perspective up to 2016. The paper by Nielsen et al. (2017) is a great starting point for understanding human migrations across the globe. The following publications deal with the regions indicated: Reich et al. (2009), Kooner et al. (2011), India; Fu et al. (2014), western Asia; Chen et al. (2009), eastern Asia; Pagani et al. (2016), Eurasia; Lazaridis et al. (2014), Europe; and Lachance et al. (2012), Africa. The Pickrell and Reich quote is from the 2014 review in *Trends in Genetics*.

Chen, J., H. Zheng, J.-X. Bei, L. Sun, W.-h. Jia, T. Li, F. Zhang, M. Seielstad, Y.-X. Zeng, X. Zhang and J. Liu. 2009. "Genetic Structure of the Han Chinese Population Revealed by Genome-Wide SNP Variation." *American Journal of Human Genetics* 85: 775–785.

Fu, Q., H. Li, P. Moorjani, F. Jay, S. M. Slepchenko, A. A. Bondarev, P. L. F. Johnson, et al. 2014. "Genome Sequence of a 45,000-Year-Old Modern Human from Western Siberia." *Nature* 514: 445–449.

Lachance, J., B. Vernot, C. C. Elbers, B. Ferwerda, A. Froment, J.-M. Bodo, G. Lema, et al. 2012. "Evolutionary History and Adaptation from High-Coverage Whole-Genome Sequences of Diverse African Hunter-Gatherers." *Cell* 150 (3): 457–469.

Lazaridis, I., N. Patterson, A. Mittnik, G. Renaud, S. Mallick, K. Kirsanow, P. H. Sudmant, et al. 2014. "Ancient Human Genomes Suggest Three Ancestral Populations for Present-Day Europeans." *Nature* 513 (7518): 409–413.

Nielsen, R., J. M. Akey, M. Jakobsson, J. K. Pritchard, S. Tishkoff, and E. Willerslev. 2017. "Tracing the Peopling of the World Through Genomics." *Nature* 541 (7637): 302–310.

Pagani, L., D. J. Lawson, E. Jagoda, A. Mörseburg, A. Eriksson, M. Mitt, F. Clemente, et al. 2016. "Genomic Analyses Inform on Migration Events During the Peopling of Eurasia." *Nature* 538 (7624): 238–242.

Pickrell, J. K., and D. Reich. 2014. "Toward a New History and Geography of Human Genes Informed by Ancient DNA." *Trends in Genetics* 30 (9): 377–389.

Reich, D., K. Thangaraj, N. Patterson, A. L. Price, and L. Singh. 2009. "Reconstructing Indian Population History." *Nature* 461 (7263): 489–494.

Slatkin, M., and F. Racimo. 2016. "Ancient DNA and Human History." *Proceedings of the National Academy of Sciences of the United States of America* 113: 6380–6387.

11. Gene Genealogies and Species Trees

A review of incongruence in different animal groups can be found in Rosenfeld, Payne, and DeSalle (2012). A discussion of tree building in reference to human population relationships can be found in DeSalle (2016) and DeSalle et al. (2017). The trees shown in this chapter are from DeSalle et al. (2017), and raw data can be obtained from the authors of that publication. The Brower, DeSalle, and Vogler (1996) paper is an accessible philosophical treatment of the "line of death." The Gower quote comes from his "Measures of Taxonomic Distance and Their Analysis."

Brower, A. V. Z., R. DeSalle, and A. Vogler. 1996. "Gene Trees, Species Trees, and Systematics: A Cladistic Perspective." *Annual Review of Ecology and Systematics* 27 (1): 423–450.

DeSalle, R. 2016. "What Do Our Genes Tell Us About Our Past?" *Journal of Anthropological Sciences* 94: 193–200.

DeSalle, R., A. Narechania, M. Zilversmit, J. Rosenfeld, and M. Tessler. 2017. "To Tree or Not to Tree *Homo sapiens*?" In *Rethinking Human Evolution*, ed. J. Schwartz, 105–125. Cambridge, MA: MIT Press.

Gower, J. C. 1972. "Measures of Taxonomic Distance and Their Analysis." In *The Assessment of Population Affinities in Man*, ed. J. S. Weiner and J. Huizinga, 1–24. Oxford: Clarendon.

Rosenfeld, J. A., A. Payne, and R. DeSalle. 2012. "Random Roots and Lineage Sorting." *Molecular Phylogenetics and Evolution* 64 (1): 12–20.

12. Clustering Humans?

Galton's classic work on the efficacy of prayer was published in 1872. Zubin's psychological clustering work was reported in 1938, and the Polynesian clustering diagram is from Driver and Kroeber (1932). The quote from Francois et al. comes from their excellent review of clustering methods in evolutionary biology published in 2010. Cavalli-Sforza and colleagues (e,g, 1993) were early users of PCA in the context of human biogeography. Price et al. (2006) described the first iteration of the EIGENSTRAT algorithm. The Jain quote is from a paper published in *Pattern Recognition Letters* (2010). References from the table in figure 12.1 are also given below.

Auton, A., K. Bryc, A. R. Boyko, K. E. Lohmueller, J. Novembre, A. Reynolds, A. Indap, et al. 2009. "Global Distribution of Genomic Diversity Underscores Rich Complex History of Continental Human Populations." *Genome Research* 19 (5): 795–803.

Biswas, S, L. B. Scheinfeldt, and J. M. Akey. 2009. "Genome-Wide Insights into the Patterns and Determinants of Fine-Scale Population Structure in Humans." *American Journal of Human Genetics* 84 (5): 641–650.

Cavalli-Sforza, L., P. Menozzi, and A. Piazza. 1993. *The History and Geography of Human Genes*. Princeton, NJ: Princeton University Press.

Cherni, L., A. J. Pakstis, S. Boussetta, S. Elkamel, S. Frigi, H. Khodjet el-Khil, A. Barton, et al. 2016. "Genetic Variation in Tunisia in the Context of Human Diversity Worldwide." *American Journal of Physical Anthropology* 161 (1): 62–71.

Driver, H. E., and A. L. Kroeber. 1932. *Quantitative Expression of Cultural Relationships.* Berkeley: University of California Press.

François, O., M. Currat, N. Ray, E. Han, L. Excoffier, and J. Novembre. 2010. "Principal Component Analysis Under Population Genetic Models of Range Expansion and Admixture." *Molecular Biology and Evolution* 27 (6): 1257–1268.

Galton, F. 1872. "Statistical Inquiries into the Efficacy of Prayer." *Fortnightly Review* 12: 125–135.

Jain, Anil K. 2010. "Data Clustering: 50 Years Beyond K-Means." *Pattern Recognition Letters* 31 (8): 651–666.

Lek, M., K. J. Karczewski, E. V. Minikel, K. E. Samocha, E. Banks, T. Fennell, A. H. O'Donnell-Luria, et al. 2016. "Analysis of Protein-Coding Genetic Variation in 60,706 Humans." *Nature* 536 (7616): 285–291.

Li, Jun Z., D. M. Absher, H. Tang, A. M. Southwick, A. M. Casto, S. Ramachandran, H. M. Cann, et al. 2008. "Worldwide Human Relationships Inferred from Genome-Wide Patterns of Variation." *Science* 319 (5866): 1100–1104.

Martin, R., and the 1000 Genomes Project Consortium. 2015. "A Global Reference for Human Genetic Variation." *Nature* 526 (7571): 68–74.

Nassir, R., K. C. Tian, P. A. White, L. M. Butler, G. Silva, R. Kittles, et al. 2009. "An Ancestry Informative Marker Set for Determining Continental Origin: Validation and Extension Using Human Genome Diversity Panels." *BMC Genetics* 10 (1): 39.

Paschou, P., E. Ziv, E. G. Burchard, S. Choudhry, W. Rodriguez-Cintron, M. W. Mahoney, and P. Drineas. 2007. "PCA-Correlated SNPs for Structure Identification in Worldwide Human Populations." *PLoS Genetics* 3 (9): e160.

Price, A. L., N. J. Patterson, R. M. Plenge, M. E. Weinblatt, N. A. Shadick, and D. Reich. 2006. "Principal Components Analysis Corrects for Stratification in Genome-Wide Association Studies." *Nature Genetics* 38 (8): 904–909.

Sneath, P. H. A., and R. R. Sokal. 1963. *Numerical Taxonomy. The Principles and Practice of Numerical Classification.* San Francisco: Freeman.

Sudmant, P. H., T. Rausch, E. J. Gardner, R. E. Handsaker, A. Abyzov, J. Huddleston, Y. Zhang, et al. 2015. "An Integrated Map of Structural Variation in 2,504 Human Genomes." *Nature* 526 (7571): 75–81.

Zubin, J. 1938. "A Technique for Measuring Like-Mindedness." *Journal of Abnormal and Social Psychology* 33 (4): 508.

13. STRUCTUREing Humans?

The Pritchard, Stephens, and Donnelly (2000) study cited in the text describes the nuances of cluster analysis. The description of the original STRUCTURE program is also in this publication. ADMIXTURE was first described by Alexander, Novembre, and Lange (2009). fineSTRUCTURE is described by Lawson et al. (2012). Rosenberg and colleagues' original study is Rosenberg et al. (2002), and the 2005 study is Ramachandran et al. (2005). Wade (2015) inappropriately cites this work.

Li et al. (2008) studied Chinese populations. The teraSTRUCTURE program was first described in Gopalan et al. (2016). The Bryc et al. (2010) reference comes from the Tishkoff lab and adds several African genomes to its population structure analysis. Edwards's "Lewontin's fallacy" paper was published in 2003. The quote from Novembre and Peter can be found in their paper published in *Current Opinion in Genetics & Development* in 2016. A modern look at Lewontin's 85/15 percent description of human diversity appears in Witherspoon et al. (2007).

Alexander, D. H., J. Novembre, and K. Lange. 2009. "Fast Model-based Estimation of Ancestry in Unrelated Individuals." *Genome Research* 19 (9): 1655–1664.

Bryc, K., A. Auton, M. R. Nelson, J. R. Oksenberg, S. L. Hauser, S. Williams, A. Froment, et al. 2010. "Genome-Wide Patterns of Population Structure and Admixture in West Africans and African Americans." *Proceedings of the National Academy of Sciences of the United States of America* 107: 786–791.

Edwards, A. W. F. 2003. "Human Genetic Diversity: Lewontin's Fallacy." *BioEssays* 25 (8): 798–801.

Gopalan, P., W. Hao, D. M. Blei, and J. D. Storey. 2016. "Scaling Probabilistic Models of Genetic Variation to Millions of Humans." *Nature Research* 48 (12): 1587–1590.

Gower, J. C. 1972. "Measures of Taxonomic Distance and Their Analysis." In *The Assessment of Population Affinities in Man*, ed. J. S. Weiner and J. Huizinga, 1–24. Oxford: Clarendon.

Lawson, D. J., G. Hellenthal, S. Myers, and D. Falush. 2012. "Inference of Population Structure Using Dense Haplotype Data." *PLoS Genetics* 8 (1): e1002453.

Li, J., D. M. Absher, H. Tang, A. M. Southwick, A. M. Casto, S. Ramachandran, H. M. Cann, et al. 2008. "Worldwide Human Relationships Inferred from Genome-Wide Patterns of Variation." *Science* 319 (5866): 1100–1104.

Novembre, J., and B. M. Peter. 2016. "Recent Advances in the Study of Fine-Scale Population Structure in Humans." *Current Opinion in Genetics & Development* 41: 98–105.

Pritchard, J. K., M. Stephens, and P. Donnelly. 2000. "Inference of Population Structure Using Multilocus Genotype Data." *Genetics* 155 (2): 945–959.

Ramachandran, S., O. Deshpande, C. C. Roseman, N. A. Rosenberg, M. W. Feldman, and L. L. Cavalli-Sforza. 2005. "Support from the Relationship of Genetic and Geographic Distance in Human Populations for a Serial Founder Effect Originating in Africa." *Proceedings of the National Academy of Sciences of the United States of America* 102: 15942–15947.

Rosenberg, N. A., J. K. Pritchard, J. L. Weber, H. M. Cann, K. K. Kidd, L. A. Zhivotovsky, and M. W. Feldman. 2002. "Genetic Structure of Human Populations." *Science* 298 (5602): 2381–2385.

Wade, N. 2015. *A Troublesome Inheritance: Genes, Race and Human History*. New York: Penguin.

Witherspoon, D. J., S. Wooding, A. R. Rogers, E. E. Marchani, W. S. Watkins, M. A. Batzer, and L. B. Jorde. 2007. "Genetic Similarities Within and Between Human Populations." *Genetics* 176 (1): 351–359.

14. Mr. Murray Loses His Bet

Charles Murray's review of Wade's 2015 book was published in the *Wall Street Journal* (Murray 2014). The quote from Wade's 2014 broadcast interview with Fareed Zakaria may be found at https://archive.org/details/CNNW_20140608_170000 _Fareed_Zakaria_GPS. The paper by "Hamilton" on taxonomy and race appeared in *Occidental Quarterly* (Hamilton 2008). Wade relied heavily on two companion studies to make his case that races exist: Rosenberg et al. (2002) and Ramachandran et al. (2005). The *Faces of America* PBS television show referred to in this chapter has a website (www.pbs.org/wnet/facesofamerica) that can be accessed to view Stephen Colbert calling himself the "black hole of white people" after reviewing his ancestry results.

Hamilton, A. [pseud.]. 2008. "Taxonomic Approaches to Race." *Occidental Quarterly* 8 (3): 11–36.

Livingstone, F. B. 1962. "On the Non-Existence of Human Races." *Current Anthropology* 3: 279–281.

Murray, C. 2014. Book review of *A Troublesome Inheritance*, by Nicholas Wade. *Wall Street Journal*, May 2, 2014. https://www.wsj.com/articles/book-review-a -troublesome-inheritance-by-nicholas-wade-1399066489).

Ramachandran, S., O. Deshpande, C. C. Roseman, N. A. Rosenberg, M. W. Feldman, and L. L. Cavalli-Sforza. 2005. "Support from the Relationship of Genetic and Geographic Distance in Human Populations for a Serial Founder Effect Originat-ing in Africa." *Proceedings of the National Academy of Sciences of the United States of America* 102: 15942–15947.

Rosenberg, N. A., J. K. Pritchard, J. L. Weber, H. M. Cann, K. K. Kidd, L. A. Zhivotovsky, and M. W. Feldman. 2002. "Genetic Structure of Human Populations." *Science* 298 (5602): 2381–2385.

Tattersall, I., and R. DeSalle. 2011. *Race? Debunking a Scientific Myth*. College Station: Texas A&M University Press.

Wade, N. 2015. *A Troublesome Inheritance: Genes, Race and Human History*. New York: Penguin.

Zakaria, F. Interview with Nicolas Wade. *Fareed Zakaria GPS*. CNN, June 8, 2014. https://archive.org/details/CNNW_20140608_170000_Fareed_Zakaria_GPS.

Epilogue: Race and Society

The sociologists Kevin Wallsten, Tatishe M. Nteta, and Lauren A. McCarthy (2015) have emphasized the relationship between racial prejudice and the reluctance to pay college athletes. Taylor Branch's statement on the matter can be found in his article in the *Atlantic* (2015). Robin Dunbar's estimated maximum number of close associates is elaborated in *Grooming, Gossip and the Evolution of Language* (1998). Stanley Garn summarized his attempts to classify populations within the human species in *Human Races* (1961). Farley (1996) and Tuch, Siegelman, and Macdonald

(1999) presented evidence that white belief in black inferiority had been waning in America for some time, and Welch and Siegelman (2011) suggested that the election of Barack Obama to the U.S. presidency had accentuated this decline. Nteta and Greenlee (2013) studied subsequent generational effects among whites. The website quote from Nteta was retrieved from https://polsci.umass.edu/news /fatherhood-race-politics-research-profile-tatishe-nteta.

Branch, T. 2015. "Toward Basic Rights for College Athletes." *Atlantic*, November 11, 2012. www.theatlantic.com/business/archive/2015/11/ncaa-taylor-branch/415389.

Dunbar, R. I. M. 1998. *Grooming, Gossip and the Evolution of Language*. Cambridge, MA: Harvard University Press.

Farley, R. 1996. *The New American Reality: Who We Are, How We Got There, Where We Are Going*. New York: Russell Sage Foundation.

Garn, S. M. 1961. *Human Races*. Springfield, IL: Thomas.

Nteta, T., and J. S. Greenlee. 2013. "A Change Is Gonna Come: Generational Membership and White Racial Attitudes in the 21st Century." *Political Psychology* 34 (6): 877–897.

Tuch, S., L. Siegelman, and J. Macdonald. 1999. "Race Relations and American Youth, 1976–1985." *Public Opinion Quarterly* 63: 109–148.

Wallsten, K., T. M. Nteta, and L. A. McCarthy. 2015. "Racial Prejudice Is Driving Opposition to Paying College Athletes. Here's the Evidence." *Washington Post*, December 30, 2015.

Welch, S., and L. Siegelman. 2011. "The 'Obama Effect' and White Racial Attitudes." *Annals of the American Academy of Political and Social Science* 634 (1): 207–220.

Index

Page numbers in italics indicate figures or tables.